ENCYCLOPAEDIA OF PHOTOSYNTHESIS

ENCYCLOPAEDIA
OF
PHOTOSYNTHESIS

Vol. 2

Udai Veer

ANMOL PUBLICATIONS PVT. LTD.
NEW DELHI - 110 002 (INDIA)

ANMOL PUBLICATIONS PVT. LTD.
H.O.: 4374/4B, Ansari Road, Darya Ganj,
New Delhi-110 002 (India)
Ph.: 23278000, 23261597
B.O.: No. 1015, Ist Main Road, BSK IIIrd Stage
IIIrd Phase, IIIrd Block,
Bangalore - 560 085 (India)
Visit us at: www.anmolpublications.com

Encyclopaedia of Photosynthesis

First Adition, 2008
ISBN 978-81-261-3474-8 (Set)

PRINTED IN INDIA

Printed at Mehra Offset Press, Delhi.

Contents

Preface

Phtosynthesis is the process, by which plants make their food in their green leaves with the help of carbon dioxide, minerals, and water; in the presence of light. In this process oxygen is given out, which is very useful in the respiration. This is the process, in which light energy is captured in green leaves for the chemical process, which makes it most important and fascinating of biological processes. By the absorption of blue light, chlorophyll emits only the useful red fluorescence. The molecule goes from its second to first excited singlet state by a radiationless process of high probability. The carbohydrate made in the process of photosynthesis is synthesised by animals and plants both. They synthesise fats and proteins from the carbohydrate; thus glucose is a basic energy source for all living organisms.

Photosynthesis fills all of the food requirements directly or indirectly. The energy stored in petroleum, natural gas and coal, has all come from the sun via photosynthesis, as does the energy in firewood, which is a major fuel in many parts of the world. All of the biological energy needs are met by the plant kingdom, either directly or through herbivorous animals. One of the carbohydrates resulting from photosynthesis is cellulose, which makes up the bulk of dry wood and other plant material. When the wood is burnt the cellulose is converted back to carbon dioxide and releases the stored energy as heat. Burning fuel is basically the same oxidation process that occurs in living bodies. It liberates the energy of stored sunlight in a useful form, and returns carbon dioxide to the atmosphere.

Energy from burning biomass is important in many parts of the world. In developing countries, firewood continues to be critical for survival. Ethanol (grain alcohol) produced from sugar and starch by fermentation, is a major automobile fuel in Brazil and is added to gasoline in some parts of the United States in order to help reduce emission of harmful pollutants.

Photosynthesis by plants removes carbon dioxide from the atmosphere and replaces it with oxygen. Thus, it tends to ameliorate the effects of carbon dioxide released by the burning of the fossil fuels. Photosynthetic pigments come in a huge variety. There are different types of chlorophyll, carotenoids and phycobilins, differing from each other, in their precise chemical structure.

Generally pigments are bound by proteins, which provide the pigment molecules with appropriate positioning and orientation, with respect to each other.

This eminent work: *Encyclopaedia of Photosynthesis* has a very realistic and systematic feature, which would be able to make a distinct impression on the study of photosynthesis. Expectedly, each and every reader will get full satisfaction by studying this book.

Editor

Photosynthetic Pigments

A photosynthetic pigment or antenna pigment is a pigment that is present in chloroplasts or photosynthetic bacteria and captures the light energy necessary for photosynthesis.

Plants

Green plants have five closely-related photosynthetic pigments (in order of increasing polarity):

- Carotene — an orange pigment
- Xanthophyll — a yellow pigment
- Chlorophyll a — a blue-green pigment
- Chlorophyll b — a yellow-green pigment
- Phaeophytin — a grey pigment

Chlorophyll a is the most common of the five, present in every plant that performs photosynthesis. The reason that there are so many pigments is that each absorbs light more efficiently in a different part of the spectrum. Chlorophyll absorbs well at a wavelength of about 400-450 nm and at 650-700 nm; chlorophyll b at 450-500 nm. Xanthophyll absorbs well at 400-530 nm. However, none of the pigments absorbs well in the green-yellow region, which is responsible for the abundant green we see in nature.

Bacteria

Like plants, the cyanobacteria use water as an electron donor for photosynthesis and therefore liberate oxygen; they also use chlorophyll as a pigment. In addition, most cyanobacteria use phycobiliproteins to capture light energy and pass it on to the chlorophylls. (Some cyanobacteria, the prochlorophytes, use chlorophyll b instead of phycobilin.) It is thought that the chloroplasts in plants and algae all evolved from cyanobacteria.

Several other groups of bacteria use the bacteriochlorophyll pigments (similar to the chlorophylls) for photosynthesis. Unlike the cyanobacteria, these bacteria do not produce oxygen; they typically use hydrogen sulphide rather than water as the electron donor. Recently, a very different pigment has been found in some marine γ-proteobacteria: proteorhodopsin. It is similar to and probably originated from bacteriorhodopsin.

Algae

Green algae, red algae and glaucophytes all use chlorophylls. Red algae and glaucophytes also use phycobiliproteins, but green algae do not.

Archaea

Photosynthesis in archaea is quite different from the systems in other domains of life. Photosynthetic archaea (the halobacteria) use the pigment bacteriorhodopsin which acts directly as a proton pump when exposed to light.

Carotenoids

Carotenoids are organic pigments that are naturally occurring in plants and some other photosynthetic organisms like algae, some types of fungus and some bacteria. There are over 600 known carotenoids; they are split into two classes, xanthophylls and carotenes. They absorb blue light.

Properties

Carotenoids belong to the category of tetraterpenoids (i.e. they contain 40 carbon atoms). Structurally they are in the form of a polyene chain which is sometimes terminated by rings.

- Carotenoids with molecules containing oxygen, such as lutein and zeaxanthin, are known as xanthophylls.
- The unoxidised (oxygen free) carotenoids such as alpha-carotene, beta-carotene and lycopene are known as carotenes. Carotenes typically contain only carbon and hydrogen.

Probably the most well-known carotenoid is the one that gives this second group its name, carotene, found in carrots and responsible for their bright orange colour. Crude palm oil, however, is the richest source of carotenoids in nature.

Their colour, ranging from pale yellow through bright orange to deep red, is directly linked to their structure. Xanthophylls are often yellow, hence their class name. The double carbon-carbon bonds interact with each other in a process called conjugation, which allows electrons in the molecule to move freely across these areas of the molecule. As the number of double bonds increases, electrons associated with conjugated systems have more room to move, and require less energy to change states. This causes the range of energies of light absorbed by the molecule to decrease. As more frequencies of light are absorbed from the short end of the visible spectrum, the compounds acquire an increasingly red appearance.

Physiological Effects

In photosynthetic organisms, carotenoids play a vital role in the photosynthetic reaction centre. They either participate in the energy-transfer process, or protect the reaction centre from auto-oxidation. In non-photosynthesising organisms, carotenoids have been linked to oxidation-preventing mechanisms.

Carotenoids have many physiological functions. Carotenoids are efficient free-radical scavengers, and they enhance the vertebrate immune system. Consequently, epidemiological studies have shown that people with high beta-carotene intake and high plasma levels of beta-carotene have a significantly reduced risk of lung cancer. But studies of supplementation with large doses of beta-carotene in smokers have shown an increase in cancer risk (possibly because excessive beta-carotene results in breakdown products that reduce plasma vitamin A and worsen the lung cell proliferation induced by smoke). Similar results have been found in other animals.

Animals are incapable of synthesising carotenoids, and must obtain them through their diet, yet they are common and often in ornamental features. For example, the pink colour of flamingos and salmon, and the red colouring of lobsters are due to carotenoids. It has been proposed that carotenoids are used in ornamental traits because, given their physiological and chemical properties, they can be used as honest indicators of individual health, and hence they can be used by animals when selecting potential mates.

The most common carotenoids include lycopene and the vitamin A precursor β-carotene. In plants, the xanthophyll lutein is the most abundant carotenoid and its role in preventing age-related eye disease is currently under investigation. Lutein and the other carotenoid pigments found in leaves are not obvious because of the presence of other pigments such as chlorophyll.

Aroma Chemicals

Products of carotenoid degradation such as ionones, damascones, and damascenones are also important fragrance chemicals that are used extensively in the perfumes and fragrance industry. Both beta-damascenone and beta-ionone although low in concentration in rose distillates are the key odour-contributing compounds in flowers. In fact, the sweet floral smells present in black tea, aged tobacco, grape, and many fruits are due to the aromatics compounds resulting from carotenoid breakdown.

Apocarotenal

Apocarotenal, or trans-beta-apo-8'-carotenal, is a carotenoid found in spinach and citrus fruits. Like other carotenoids, apocarotenal plays a role as a precursor of vitamin A, even though it has 50 per cent less pro-vitamin A activity than beta-carotene.

Apocarotenal has an orange to orange-red colour and is used in foods, pharmaceuticals and cosmetic products. Depending on the product forms, apocarotenal is used in fat based food (margarine, sauces, salad dressing), beverages, dairy products and sweets.

Astaxanthin

Astaxanthin (pronounced as-tuh-zan'-thin) is a carotenoid. It belongs to a larger class of phytochemicals known as terpenes. It is classified as a xanthophyll, which means "yellow leaves". Like many carotenoids, it is a colourful, fat/oil-soluble pigment. Astaxanthin can be found in microalgae, yeast, salmon, trout, krill, shrimp, crayfish, crustaceans, and the feathers of some birds. Professor Basil Weedon was the first to map the structures of astaxanthin.

Astaxanthin, unlike some carotenoids, does not convert to Vitamin A (retinol) in the human body. Too much Vitamin A is toxic for a human, but astaxanthin is not. However, it is a powerful antioxidant; it is 10 times more capable than other carotenoids.

While astaxanthin is a natural nutritional component, it can be found as a food supplement. The supplement is intended for human, animal, and aquaculture consumption. The commercial production of astaxanthin comes from both natural and synthetic sources.

The FDA approved astaxanthin as a food colouring (or colour additive) for specific uses in animal and fish foods. The European Union (actually European Commission) considers it food dye.

Natural Sources

The following sources are being used for the commercial production of astaxanthin.

- Haematococcus pluvialis (microalgae)
- Xanthophyllomyces dendrorhous, formerly Phaffia rhodozyma (yeast)
- Euphausia superba (krill)
- Pandalus borealis (shrimp)

As a natural source, the following can be found in nature (or a production facility) with the approximate concentrations listed.

Amounts of astaxanthin found in food

Natural Astaxanthin Sources	*Astaxanthin concentration(ppm)*
Salmonids	~ 5
Plankton	~ 60
Krill	~ 120.
Arctic shrimp	~ 1200
Phaffia Yeast	~ 8000
Haematococcus pluvialis	~40,000

With that, each source has its own production issues.

Currently, the primary natural source for astaxanthin is *Haematococcus pluvialis* (microalgae). It seems to accumulate the highest levels of astaxanthin in nature; commercially more than 40 g of astaxanthin per kilo of dry biomass. It has the advantage of the population doubling every week, which means scaling up is not an issue; simply open another plant. However, it does require some expertise among the personnel and that might be its downfall.

Canthaxanthin

Canthaxanthin is a carotenoid. It belongs to a larger class of phytochemicals known as *terpenes*. The chemical formula is $C_{40}H_{52}O_2$. It has E number E161g.

It was first isolated in edible mushrooms. It has also been found in green algae, bacteria, crustacea, and fishes such as carp, golden mullet, seabream and trush wrasse.

In the United Kingdom, canthaxanthin is allowed by law to be added to poultry feed, some sausages mainly exported to France, and trout and salmon feed. In addition, the European Union has set the limit at 80mg/Kg of foodstuffing.

Carotene

The term carotene is used for several related substances having the formula $C_{40}H_{56}$. Carotene is an orange photosynthetic pigment important for photosynthesis. It is responsible for the orange colour of the carrot and many other fruits and vegetables. It contributes to photosynthesis by transmitting the light energy it absorbs to chlorophyll.

Chemically, carotene is a terpene, synthesised biochemically from eight isoprene units. It comes in two primary forms designated by characters from the Greek alphabet: alpha-carotene (a-carotene) and beta-carotene (a-carotene). Gamma, delta and epsilon also exist. Beta-carotene is composed of two retinyl groups, and is broken down in the mucosa of the small intestine by beta-carotene dioxygenase to retinol, a form of vitamin A. Carotene can be stored in the liver and converted to vitamin A as needed, thus making it a provitamin.

Crocetin

Crocetin is a natural carotenoid dicarboxylic acid that is found in the crocus flower. It forms brick red crystals with a melting point of 285 °C. The chemical structure of crocetin is the central core of crocin, the compound responsible for the colour of saffron.

Crocin

Crocin is a natural carotenoid chemical compound that is found in the flowers crocus and gardenia. It is the diester formed from the disaccharide gentiobiose and the dicarboxylic acid crocetin. It has a deep red colour and forms crystals with a melting point of 186°C. When dissolved in water, it forms an orange solution.

Crocin is the chemical ingredient primarily responsible for the colour of saffron. It has also been shown to be a potent antioxidant.

Cryptoxanthin

Cryptoxanthin is a natural carotenoid pigment. It has been isolated from a variety of sources including the petals and flowers of plants in the genus Physalis, orange rind, papaya, egg yolk, butter, and bovine blood serum.

Structurally, cryptoxanthin is closely related to beta-carotene, with only the addition of a hydroxyl group. It is a member of the class of carotenoids known as xanthophylls.

In a pure form, cryptoxanthin is a red crystalline solid with a metallic luster. It is freely soluble in chloroform, benzene, pyridine and carbon disulphide.

Fucoxanthin

Fucoxanthin is a carotenoid, with formula $C_{40}H_{60}O_6$. It is found as an accessory pigment in the chloroplasts of brown algae and most other heterokonts, giving them a brown or olive-green colour. Fucoxanthin absorbs light primarily in the blue-green to yellow-green part of the visible spectrum, peaking at around 510-525 nm by various estimates and absorbing significantly in the range of 450 to 540 nm.

Some metabolic and nutritional studies carried at Hokkaido University have demonstrated that brown seaweed (rich in fucoxanthin) promotes fat burning within fat cells surrounding internal organs. In addition, fucoxanthin has strong anti-diabetes effects and, by a separate biological mechanism, it appears to kill prostate cancer cells in culture.

Lutein

Lutein (LOO-teen) (from Latin *lutea* meaning "yellow") is one of over 600 known naturally occurring carotenoids. Found in green leafy vegetables such as spinach and kale, lutein is employed by organisms as an antioxidant and for blue light absorption. Lutein is covalently bound to one or more fatty acids present in some fruits and flowers, notably marigolds (Tagetes). Saponification of lutein esters yields lutein in approximately a 2:1 weight-to-weight conversion.

Lutein is a lipophilic molecule and is generally insoluble in water. The presence of the long chromophore of conjugated double bonds (polyene chain) provides the distinctive light-absorbing properties. The polyene chain is susceptible to oxidative degradation by light or heat and is chemically unstable in acids. The principal natural stereoisomer of lutein is (3R,3′R,6′R)-beta, epsilon-Carotene-3,3'-diol.

This xanthophyll, like its sister compound zeaxanthin, has primarily been used as a natural colourant due to its orange-red colour. Lutein absorbs blue light and therefore appears yellow at low concentrations and orange-red at high concentrations. Lutein was traditionally used in chicken feed to provide the yellow colour of broiler chicken skin. Polled consumers viewed yellow chicken skin more favourably than white chicken skin. Such lutein fortification also results in a darker yellow egg yolk. Today the colouring of the egg yolk has become the primary reason for feed fortification. Lutein is not used as a colourant in other foods due to its limited stability, especially in the presence of other dyes.

Lycopene

Lycopene is a bright red carotenoid pigment, a phytochemical found in tomatoes and other red fruits. Lycopene is the most common carotenoid in the human body and is one of the most potent carotenoid antioxidants. Its name is derived from the tomato's species classification, *Solanum lycopersicum* (formerly *Lycopersicon esculentum*).

Lycopene is a terpene assembled from 8 isoprene units.

The colour of lycopene is due to its many conjugated carbon double bonds. Each double bond reduces the energy required for electrons to transition to higher energy states, allowing the molecule to absorb visible light of progressively longer wavelengths. Lycopene absorbs most of the visible spectrum, so it appears red.

If lycopene is oxidised (for example, by reacting with bleaches or acids), the double bonds between carbon atoms will be broken, cleaving the molecule into smaller molecules each double-bonded to an oxygen atom. Although C=O bonds are also chromophoric, the much shorter molecules are unable to absorb enough light to appear colourful. A similar effect occurs if lycopene is reduced; reduction may saturate (convert the double bonds to single bonds) the lycopene molecule, diminishing its ability to absorb light.

Tretinoin

Tretinoin is the acid form of vitamin A and so also known as all-trans retinoic acid or ATRA. It is a drug commonly used to treat acne vulgaris and keratosis pilaris. It is available as a cream or gel (brand name Retin-A). It is also used to treat acute promyelocytic leukaemia (APML).

Xanthophyll

Xanthophylls (originally phylloxanthins) are yellow pigments of oxycarotenoid type, from the carotenoid group. Their chemical structure is $C_{40}H_{54}(OH)_2$. They are found in the leaves of most plants and are synthesised within the plastids. They are involved in photosynthesis along with green chlorophyll, which typically covers up the yellow except in autumn, when the chlorophyll is denatured by the cold.

In plants, xanthophylls are considered accessory pigments, along with anthocyanins, carotenes, and sometimes phycobiliproteins. Xanthophylls, along with carotenic pigments are seen when leaves turn orange in the autumn season.

Animals cannot produce xanthophylls, and thus xanthophylls found in animals (e.g. in the eye) come from their food intake. The yellow colour of chicken egg yolks also comes from ingested xanthophylls. Xanthophylls are oxidised derivatives of carotenes. They contain hydroxyl groups and are more polar; therefore, they are the pigments that will travel the furthest in paper chromatography.

The group of xanthophylls is composed of lutein, zeaxanthin, and α- and β-cryptoxanthin.

Xanthophyll Cycle

The xanthophyll cycle involves conversions of pigments from a non-energy-quenching form to energy-quenching forms. This is a way to reduce the absorption cross-section of the light harvesting antenna, and thus to reduce the amount of energy that reaches the photosynthetic reaction centres. Reducing the light harvesting antenna is one of the main ways of protecting against photoinhibition and changes in the xanthophyll cycling takes place on a time scale of minutes to hours.

In higher plants there are three carotenoid pigments that are active in the xanthophyll cycle: violoxanthin, antheraxanthin and zeaxanthin. During light stress violoxanthin is converted to antheraxanthin and zeaxanthin, which functions as photoprotective pigments. In diatoms and dinoflagellates the xanthophyll cycle consists of the pigment diadinoxanthin, which is transformed into diatoxanthin (diatoms) or dinoxanthin (dinoflagellates), at high light.

Zeaxanthin

Zeaxanthin is one of the two carotenoids contained within the retina. Within the central macula, zeaxanthin is the dominant component, whereas in the peripheral retina, lutein predominates. Lutein and zeaxanthin have identical chemical formulas and are isomers, but they are not stereoisomers. The main difference between them is in the location of a double bond in one of the end rings. This difference gives lutein three chiral centres whereas zeaxanthin has two. As a food additive, zeaxanthin is a food dye with E number E161h.

Allophycocyanin

Allophycocyanin is a protein from the light-harvesting phycobiliprotein family, along with phycocyanin, phycoerythrin and phycoerythrocyanin. It is an accessory pigment to chlorophyll. All phycobiliproteins are water soluble and therefore cannot exist within the membrane like carotenoids, but aggregate forming clusters that adhere to the membrane called phycobilisomes. Allophycocyanin absorbs and emits red light (650 & 660 nm max, respectively), and is readily found in Cyanobacteria (also called blue-green algae), and red algae. Phycobilin pigments have fluorescent properties that are used in immunoassay kits. In flow cytometry, it is often abbreviated APC.

Bacteriochlorophyll

Bacteriochlorophylls are photosynthetic pigments that occur in various phototrophic bacteria. They are related to chlorophylls, which are the primary pigments in plants, algae, and cyanobacteria. Groups that contain bacteriochlorophyll conduct photosynthesis, but do not produce oxygen. They use wavelengths of light not absorbed by plants. Different groups contain different types of bacteriochlorophyll:

Bacteriochlorophyll *a*	Purple bacteria
Bacteriochlorophyll *b*	Purple bacteria
Bacteriochlorophyll *c*	Green sulphur bacteria, Chloroflexi
Bacteriochlorophyll *d*	Green sulphur bacteria
Bacteriochlorophyll *e*	Green sulphur bacteria
Bacteriochlorophyll *g*	Heliobacteria

Bacteriochlorophylls *c* and *d* are chlorins, with one reduced pyrrole ring, and the others are bacteriochlorins, with two.

Bacteriorhodopsin

Bacteriorhodopsin is a photosynthetic pigment used by archaea, most notably halobacteria. It acts as a proton pump, i.e. it captures light energy and uses it to move protons across the membrane out of the cell. The resulting proton gradient is subsequently converted into chemical energy.

Bacteriorhodopsin is an integral membrane protein usually found in two-dimensional crystalline patches known as "purple membrane", which can occupy up to nearly 50 per cent of the surface area of the archaeal cell. The repeating element of the hexagonal lattice is composed of three identical protein chains, each rotated by 120 degrees relative to the others. Each chain has seven transmembrane alpha helices and contains one molecule of retinal buried deep within, the typical structure for retinylidene proteins. It is the retinal molecule that changes its conformation when absorbing a photon, resulting in a conformational change of the surrounding protein and the proton pumping action.

The bacteriorhodopsin molecule is purple and is most efficient at absorbing green light (wavelength 500-650 nm, with the absorption maximum at 568 nm).

The three-dimensional tertiary structure of bacteriorhodopsin resembles that of vertebrate rhodopsins, the pigments that sense light in the retina. Rhodopsins also contain retinal, however the functions of rhodopsin and bacteriorhodopsin are different and there is no homology of their amino acid sequences. Both rhodopsin and bacteriorhodopsin belong to the 7TM receptor family of proteins, but rhodopsin is a G protein coupled receptor and bacteriorhodopsin is not. In the first use of electron crystallography to obtain an atomic-level protein structure, the structure of bacteriorhodopsin was resolved in 1990. It was then used as a template to build models of other G protein-coupled receptors before crystallographic structures were also available for these proteins.

Many molecules have homology to bacteriorhodopsin, including the light-driven chloride pump halorhodopsin (for whom the crystal structure is also known), and some directly light-activated channels like channelrhodopsin.

All other photosynthetic systems in bacteria, algae and plants use chlorophylls or bacteriochlorophylls rather than bacteriorhodopsin. These also produce a proton gradient, but in a quite different and more indirect way involving an electron transfer chain consisting of several other proteins. Furthermore, chlorophylls are aided in capturing light energy by other pigments known as "antennas"; these are not present in bacteriorhodopsin based systems. Lastly, chlorophyll-based photosynthesis is coupled to carbon fixation (the incorporation of carbon dioxide into larger organic molecules); this is not true for bacteriorhodopsin-based system. It is thus likely that photosynthesis independently evolved at least twice, once in bacteria and once in archaea.

Chlorophyll

Chlorophyll is a green photosynthetic pigment found in most plants, algae, and cyanobacteria. Its name is derived from ancient Greek: chloros = green and phyllon = leaf. Chlorophyll absorbs most strongly in the blue and red but poorly in the green portions of the electromagnetic spectrum, hence the green colour of chlorophyll-containing tissues like plant leaves.

Chlorophyll and Photosynthesis

Chlorophyll is vital for photosynthesis, which helps plants obtain energy from light. Chlorophyll molecules are specifically arranged in and around pigment protein complexes called photosystems, which are embedded in the thylakoid membranes of chloroplasts. In these complexes, chlorophyll serves two primary functions. The function of the vast majority of chlorophyll (up to several hundred per photosystem) is to absorb light and transfer that light energy by resonance energy transfer to a specific chlorophyll pair in the reaction centre of the photosystems. Because of chlorophyll's selectivity regarding the wavelength of light it absorbs, areas of a leaf containing the molecule will appear green.

When a leaf was tested using iodine, only the green areas were shown as positive for starch, meaning that photosynthesis will not occur without chlorophyll. Photosystem II and Photosystem I have their own distinct reaction centre chlorophylls, named P680 and P700, respectively. These pigments are named after the wavelength (in nanometres) of their red-peak absorption maximum. The identity, function and spectral properties of the types of chlorophyll in each photosystem are distinct and determined by each other and the protein structure surrounding them. Once extracted from the protein into a solvent (such as acetone or methanol), these chlorophyll pigments can be separated in a simple paper chromatography experiment, and, based on the number of polar groups between chlorophyll a and chlorphyll b, will chemically separate out on the paper.

The function of the reaction centre chlorophyll is to use the energy absorbed by and transferred to it from the other chlorophyll pigments in the photosystems to undergo a charge separation, a specific redox reaction in which the chlorophyll donates an electron into a series of molecular intermediates called an electron transport chain. The charged reaction centre chlorophyll ($P680^+$) is then reduced back to its ground state by accepting an electron. In Photosystem II, the electron which reduces $P680^+$ ultimately comes from the oxidation of water into O_2 and H^+ through several intermediates.

This reaction is how photosynthetic organisms like plants produce O_2 gas, and is the source for practically all the O_2 in earth's atmosphere. Photosystem I typically works in series with Photosystem II, thus the $P700^+$ of Photosystem I is usually reduced, via many intermediates in the thylakoid membrane, by electrons ultimately from Photosystem II. Electron transfer reactions in the thylakoid membranes are complex, however, and the source of electrons used to reduce $P700^+$ can vary. The electron flow produced by the reaction centre chlorophyll pigments is used to shuttle H^+ ions across the thylakoid membrane, setting up a chemiosmotic potential mainly used to produce ATP chemical energy, and those electrons ultimately reduce $NADP^+$ to NADPH a universal reductant used to reduce CO_2 into sugars as well as for other biosynthetic reductions.

Absorbance spectra of free chlorophyll *a* (green) and *b* (red) in a solvent. The spectra of chlorophyll molecules are slightly modified *in vivo* depending on specific pigment-protein interactions. Reaction centre chlorophyll-protein complexes are capable of directly absorbing light and performing charge separation events without other chlorophyll pigments, but the absorption cross-section (the likelihood of absorbing a photon under a given light intensity) is small. Thus, the remaining chlorophylls in the photosystem and antenna pigment protein complexes associated with the photosystems all cooperatively absorb and funnel light energy to the reaction centre.

Besides chlorophyll *a*, there are other pigments, called accessory pigments, which occur in these pigment-protein antenna complexes. They include other forms of chlorophyll, such as chlorophyll *b* in green algal and higher plant antennae, while other algae may contain chlorophyll *c* or *d*. In addition, there are many non-chlorophyll accessory pigments, such as carotenoids or phycobiliproteins which also absorb light and transfer that light energy to the photosystem chlorophylls.

Some of these accessory pigments, particularly the carotenoids, also serve to absorb and dissipate excess light energy, or work as antioxidants. The large, physically associated group of chlorophylls and other accessory pigments is sometimes referred to as a *pigment bed*, though this term is losing prominence with the advent of detailed knowledge of the structural organisation of the photosystem and antenna complexes.

The different chlorophyll and non-chlorophyll pigments associated with the photosystems all have different spectra, either because the spectra of the different chlorophyll pigments are modified by their local protein environment, or because the accessory pigments have intrinsically different absorption spectra from chlorophyll.

The net result is that, *in vivo* the total absorption spectrum is broadened and flattened such that a wider range of red, orange, yellow and blue light can be absorbed by plants and algae. Most photosynthetic organisms do not have pigments which absorb green light well, thus most remaining light under leaf canopies in forests or under water with abundant plankton is green, a spectral effect called the "green window". Some organisms, such as cyanobacteria and red algae, contain accessory phycobiliproteins that can absorb green light relatively well and thus they can exploit the little remaining green light in these habitats.

An easy experiment can show how chlorophyll is needed for photosynthesis. Some plants have variegated leaves with white and green areas. Take such a plant and put it in the dark for a few days. Take a leaf and make a "mask" to cover it out of aluminium foil. Cut two holes (one round, one square) in the foil such that one hole exposes a green part of the leaf and the other a white part of the leaf. Place the leaf in the light for an hour or so. If the leaf is exposed to a solution of iodine the green part of the leaf exposed to the light will show up black and the white part will not stain black. Furthermore, the black patch will have the same shape as the hole that you had cut in the foil. The iodine-stained starch only piles up in areas of the leaf that were green, showing that only those areas are photosynthetic. This proves that photosynthesis doesn't occur in the areas where there was no chlorophyll.

Chemical Structure

Chlorophyll is a chlorin pigment, which is structurally similar to and produced through the same metabolic pathway as other porphyrin pigments such as haem. At the centre of the chlorin ring is a magnesium ion. The chlorin ring can have several different side chains, usually including a long phytol chain. There are a few different forms that occur naturally:

	Chlorophyll a	*Chlorophyll b*	*Chlorophyll c1*	*Chlorophyll c2*	*Chlorophyll d*
Molecular formula	$C_{55}H_{72}O_5N_4Mg$	$C_{55}H_{70}O_6N_4Mg$	$C_{35}H_{30}O_5N_4Mg$	$C_{35}H_{28}O_5N_4Mg$	$C_{54}H_{70}O_6N_4Mg$
C3 group	$-CH{=}CH_2$	$-CH{=}CH_2$	$-CH{=}CH_2$	$-CH{=}CH_2$	-CHO
C7 group	$-CH_3$	-CHO	$-CH_3$	$-CH_3$	$-CH_3$
C8 group	$-CH_2CH_3$	$-CH_2CH_3$	$-CH_2CH_3$	$-CH{=}CH_2$	$-CH_2CH_3$
C17 group	$-CH_2CH_2COO-$ Phytyl	$-CH_2CH_2COO-$ Phytyl	-CH=CHCOOH	-CH=CHCOOH Phytyl	$-CH_2CH_2COO-$
C17-C18 bond	Single	Single	Double	Double	Single
Occurrence	Universal	Mostly plants	Various algae	Various algae	cyanobacteria

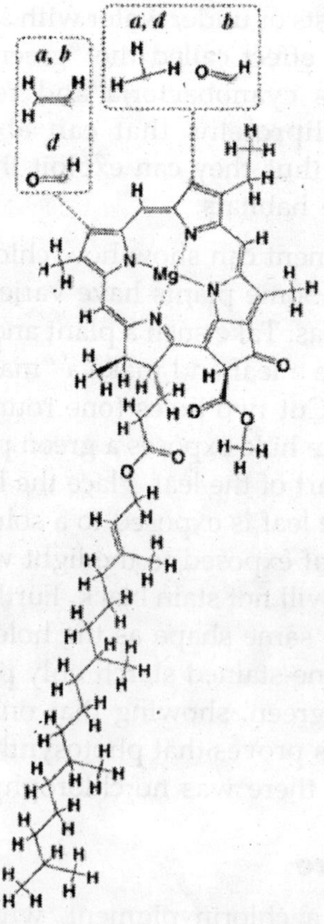

***Figure: Common structure of chlorophyll* a, b *and* d**

Figure: Common structure of chlorophyll c1, and c2

Empirical Evidence for the Importance of Chlorophyll in Photosynthesis

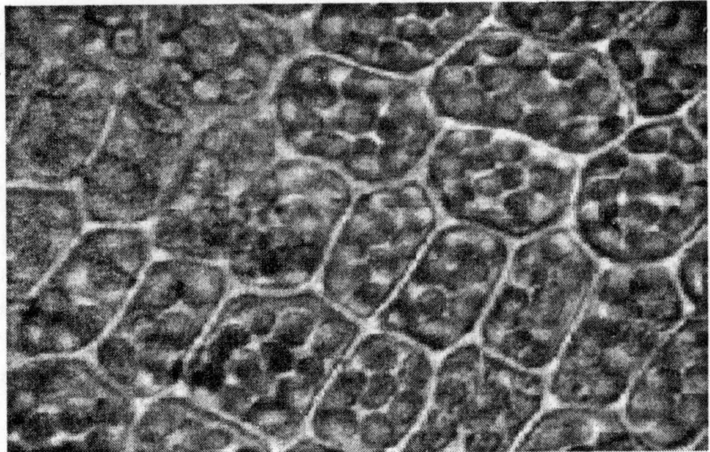

Chlorophyll is found in high concentrations in chloroplasts of plant cells.

A simple experiment can show how chlorophyll is associated with photosynthesis. After destarching a leaf from a variegated plant and exposing it to light for several hours, starch can be seen to rapidly accumulate again by staining with iodine solution. Variegated leaves have green areas that contain chlorophyll and white areas that do not. The iodine-stained starch only accumulates in regions of the leaf that were green and therefore contained chlorophyll. This shows that photosynthesis does not occur in areas where chlorophyll is absent, and constitutes evidence that the presence of chlorophyll is a requirement for photosynthesis.

Halorhodopsin

Halorhodopsin is a light-driven ion pump, specific for chloride ions, and found in phylogenetically ancient bacteria (archaea), known as halobacteria. It is a seven-transmembrane protein of the retinylidene protein family, homologous to the light-driven proton pump bacteriorhodopsin, and similar in tertiary structure (but not primary sequence structure) to vertebrate rhodopsins, the pigments that sense light in the retina. Halorhodopsin also shares sequence similarity to channelrhodopsin, a light-driven ion channel. Halorhodopsin contains the essential light-isomerisable vitamin A derivative all-trans-retinal. Due to the intense attention on solving the structure and function of this molecule, halorhodopsin is one of the few membrane proteins whose crystal structure is known.

Halorhodopsin isoforms can be found in multiple species of halobacteria, including H. salinarium, H. halobium, and N. pharaonis. Much ongoing research is exploring these differences, and using them to parse apart the photocycle and pump properties. After bacteriorhodopsin, halorhodopsin may be the best type I (microbial) opsin studied. Peak absorbance of the halorhodopsin retinal complex is about 570 nm.

Recently, attention has turned towards bioengineering uses for halorhodopsin. Just as the blue-light activated ion channel channelrhodopsin-2 opens up the ability to activate excitable cells (such as neurons, muscle cells, pancreatic cells, and immune cells) with brief pulses of blue light, halorhodopsin opens up the ability to silence excitable cells with brief pulses of yellow light. Thus halorhodopsin and channelrhodopsin together enable multiple-colour optical activation, silencing, and desynchronisation of neural activity, creating a powerful neuroengineering toolbox.

Phaeophytin

Phaeophytin is a grey pigment that appears in chromatographs of photosynthetic pigments. Phaeophytin has the same structure as chlorophyll; excluding the central magnesium ion, which is replaced by two protons. The Rf value for Phaeophytin is 0.83.

Photopigment

Photopigments are unstable pigments that undergo a physical or chemical change in the presence of light. The term is generally applied to the non-protein chromophore moiety of photosensitive chromoproteins, such as the pigments involved in photosynthesis and photoreception. In medical terminology, "photopigment" commonly refers to the photoreceptor proteins of the retina.

Phytochrome

Phytochrome is a photoreceptor, a pigment that plants use to detect light. It is sensitive to light in the red and far-red region of the visible spectrum. Many flowering plants use it to regulate the time of flowering based on the length of day and night (photoperiodism) and to set circadian rhythms. It also regulates other responses including the germination of seeds, elongation of seedlings, the size, shape and number of leaves, the synthesis of chlorophyll, and the straightening of the epicotyl or hypocotyl hook of dicot seedlings.

Biochemically, phytochrome is a protein with a bilin chromophore.

Phytochrome has been found in most plants including all

higher plants; very similar molecules have been found in several bacteria. A fragment of a bacterial phytochrome now has a solved three-dimensional protein structure.

Other plant photoreceptors include cryptochromes and phototropins, which are sensitive to light in the blue and ultraviolet regions of the spectrum.

Isoforms or States

Phytochromes are characterised by a red/far-red photochromicity. Photochromic pigments change their "colour" (spectral absorbance properties) upon light absorption. In the case of phytochrome the ground state is P_r, the $_r$ indicating that it absorbs red light particularly strongly. The absorbance maximum is a sharp peak 650 -670 nm, so concentrated phytochrome solutions look turquoise-blue to the human eye. But once a red photon has been absorbed, the pigment undergoes a rapid conformational change to form the P_{fr} state. Here $_{fr}$ indicates that now not red but far-red (also called near infrared; 705 - 740 nm) is preferentially absorbed. This shift in absorbance is apparent to the human eye as a more greenish colour. This image of Cph1 phytochrome shows the P_r ground state on the left and predominantly P_{fr} on the right. In plants at least P_{fr} is the physiologically active or "signalling" state. When P_{fr} absorbs far-red light it is converted back to P_r. Hence, red light makes P_{fr}, far-red light makes P_r. In plants at least P_{fr} is the physiologically active or "signalling" state. Since daylight contains a lot of red light, during the day phytochrome is mostly converted to P_{fr}. At night, phytochrome will slowly convert back to the P_r form. Treatment with far-red light will also convert P_{fr} back to P_r. Since plants use red light for photosynthesis, and reflect and transmit far-red light, the shade of other plants also can make P_{fr} into P_r, triggering a response called shade avoidance. In most plants, a suitable concentration of P_{fr} stimulates or inhibits physiological processes.

Since both the ground state P_r and excited state P_{fr} are unusually stable (P_{fr} has a half-life of hours or days) the quantum nature of this transition was not immediately recognised. These two forms are therefore commonly (though technically incorrectly) referred to as isoforms.

Biochemistry

Chemically, phytochrome consists of a *chromophore*, a single bilin molecule consisting of an open chain of four pyrrole rings, bonded to the protein moeity. It is the chromophore that absorbs light, and as a result changes the conformation of bilin and subsequently that of the attached protein, changing it from one state or isoform to the other.

The phytochrome chromophore is usually phytochromobilin, and is closely related to phycocyanobilin (the chromophore of the phycobiliproteins used by cyanobacteria and red algae to capture light for photosynthesis) and to the bile pigment bilirubin (whose structure is also affected by light exposure, a fact exploited in the phototherapy of jaundiced new-borns). The term "bili" in all these names refers to bile. Bilins are derived from the closed tetrapyrrole ring of haem by an oxidative reaction catalysed by haem oxygenase to yield their characteristic open chain. Chlorophyll too is derived from haem. In contrast to bilins, haem and chlorophyll carry a metal atom in the centre of the ring, iron or magnesium, respectively.

The P_{fr} state passes on a signal to other biological systems in the cell, such as the mechanisms responsible for gene expression. Although this mechanism is almost certainly a biochemical process, it is still the subject of much debate. It is known that although phytochromes are synthesised in the cytosol and the P_r form is localised there, the P_{fr} form, when generated by light illumination, is translocated to the cell nucleus. This implies a role of phytochrome in controlling gene expression, and many genes are known to be regulated by phytochrome, but the exact mechanism has still to be fully discovered. It has been proposed that phytochrome, in the P_{fr} form, may act as a kinase, and it has been demonstrated that phytochrome in the P_{fr} form can interact directly with transcription factors.

Discovery

The phytochrome pigment was discovered by Sterling Hendricks and Harry Borthwick at the USDA-ARS Beltsville Agricultural Research Centre in Maryland during a period from the late 1940s to the early 1960s. Using a spectrograph built from borrowed and war-surplus parts, they discovered that red light was very effective for promoting germination or triggering flowering responses. The red light responses were reversible by far-red light, indicating the presence of a photoreversible pigment.

The phytochrome pigment was identified using a spectrophotometer in 1959 by biophysicist Warren Butler and biochemist Harold Siegelman. Butler was also responsible for the name, phytochrome.

In 1983 the laboratories of Peter Quail and Clark Lagarias reported the chemical purification of the intact phytochrome molecule, and in 1985 the first phytochrome gene sequence was published by Howard Hershey and Peter Quail. By 1989, molecular genetics and work with monoclonal antibodies that more than one type of phytochrome existed; for example, the pea plant was shown to have at least two phytochrome types (then called type I (found predominantly in dar-grown seedlings) and type II (predominant in green plants)). It is now known by genome sequencing that Arabidopsis has exactly five phytochrome genes (PHYA - E) but that rice has only three (PHYA - C).

While this probably represents the condition in several di- and monocotyledonous plants, many plants are polyploid. Hence maize, for example, has six phytochromes - phyA1, phyA2, phyB1, phyB2, phyC1 and phyC2. While all these phytochromes have significantly different protein components, they all use phytochromobilin as their light-absorbing chromophore. In the late 1980s, the Vierstra lab showed that phyA is degraded by the ubiquitin system, the first identified natural target of the system to be identified in eukaryotes.

In 1996 a gene in the newly sequenced genome of the cyanobacterium *Synechocystis* was noticed to have a weak similarity to those of plant phytochromes. Jon Hughes in Berlin and Clark Lagarias at UC Davis subsequently showed that this gene indeed encoded a bona-fide phytochrome (named Cph1) in the sense that it is a red/far-red reversible chromoprotein. Presumably plant phytochromes are derived from an ancestral cyanobacterial phytochrome, perhaps by gene migration from the chloroplast to the nucleus. Subsequently phytochromes have been found in other prokaryotes including *Deinococcus radiodurans* and *Agrobacterium tumefaciens*. In *Deinococcus* phytochrome regulates the production of light-protective pigments, however in *Synechocystis* and *Agrobacterium* the biological function of these pigments is still unknown.

In 2005, the Vierstra and Forest labs at the University of Wisconsin published a three-dimensional structure of the photosensory domain of Deinococcus phytochrome. This breakthrough paper revealed that the protein chain forms a knot—a highly unusual structure for a protein.

Genetic Engineering

Around 1989, several laboratories were successful in producing *transgenic plants*, which produced elevated amounts of different phytochromes (*overexpression*). In all cases the resulting plants had conspicuously short stems and dark green leaves. Harry Smith and co-workers at Leicester University in England showed that by increasing the expression level of phytochrome A (which responds to far-red light) shade avoidance responses can be altered. As a result, plants can expend less energy on growing as tall as possible and have more resources for growing seeds and expanding their root systems.

This could have many practical benefits: for example, grass blades that would grow more slowly than regular grass would not require mowing as frequently, or crop plants might transfer more energy to the grain instead of growing taller.

Phycobilin

Phycobilins are chromophores (light-capturing molecules) found in cyanobacteria and in the chloroplasts of red algae, glaucophytes and some cryptomonads (though not in green algae and higher plants). They are unique among the photosynthetic pigments in that they are bonded to certain water-soluble proteins, known as phycobiliproteins. Phycobiliproteins then pass the light energy to chlorophylls for photosynthesis.

The phycobilins are especially efficient at absorbing red, orange, yellow and green light, wavelengths which are not well absorbed by chlorophyll *a*. Organisms growing in shallow waters tend to contain phycobilins that can capture yellow/red light, while those at greater depth often contain more of the phycobilins that can capture green light, which is relatively more abundant there.

Types

There are four types of phycobilins:

1. Phycoerythrobilin which is red
2. Phycourobilin which is orange
3. Phycoviolobilin (also known as phycobiliviolin) found in phycoerythrocyanin and
4. Phycocyanobilin (also known as phycobiliverdin which is blue.

They can be found in different combinations attached to phycobiliproteins to confer specific spectroscopic properties.

Properties

Chemically, phycobilins consist of an open chain of four pyrrole rings (*tetrapyrrole*) and are structurally similar to the bile pigment bilirubin. (Bilirubin's structure is also affected by light, a fact used for the phototherapy of jaundiced new-borns.) Phycobilins are also closely related to the chromophores of the light-detecting plant pigment phytochrome which also consist of an open chain of four pyrroles. Chlorophylls are composed of four pyrroles as well, but there they are arranged in a ring and contain a metal atom in the centre.

The phycobilins fluoresce at a particular wavelength, and are therefore often used in research as chemical tags, e.g. by binding phycobiliproteins to antibodies in a technique known as immunofluorescence.

Phycobiliprotein

Phycobiliproteins are water-soluble proteins present in cyanobacteria and certain algae (rhodophytes, cryptomonads, glaucocystophytes) that capture light energy which is then passed on to chlorophylls during photosynthesis. Phycobiliproteins are formed of a complex between proteins and covalently bound phycobilins that act as chromophores (the light-capturing part). They are most important constituents of the phycobilisomes.

The major phycobiliproteins are :

- Phycoerythrin
- Phycocyanin
- Allophycocyanin
- Phycoerythrocyanin

Phycocyanin

Phycocyanin is a protein from the light-harvesting phycobiliprotein family, along with allophycocyanin and phycoerythrin. It is an accessory pigment to chlorophyll. All phycobiliproteins are water soluble and therefore cannot exist within the membrane like carotenoids, but aggregate forming clusters that adhere to the membrane called phycobilisomes. Phycocyanin absorbs orange and red light and emits blue light. It is found in Cyanobacteria (previously called blue-green algae). Phycobiliproteins have fluorescent properties that are used in immunoassay kits. Phycocyanin is from the Greek *phyco* meaning "algae" and *cyanin* is from the English word "cyan."

Phycoerythrin

Phycoerythrin is a red protein from the light-harvesting phycobiliprotein family, isolated from cyanobacteria, red algae and cryptomonads. A strong absorption peak exists at about 566 nm, and a strong emission peak exists at 575 ± 10 nm. (i.e. phycoerythrin absorbs green light and emits red light.)

Like all phycobiliproteins, phycoerythrin contains chromophores called phycobilins (e.g. *phycoerythrobilin, phycourobilin*).

In some photosynthetic organisms, phycoerythrin is an accessory pigment to the main light-absorbing chlorophyll pigments responsible for photosynthesis. The light energy is captured by phycoerythrin and is then passed on to chlorophyll.

R-Phycoerythrin is useful in the laboratory as a fluorescence-based indicator for the presence of cyanobacteria and for labelling antibodies in a technique called immunofluorescence, among other applications. There are also other types of phycoerythrins, such as B-Phycoerythrin, which has a slightly different spectral characteristic. B-Phycoerythrin abosorbs strongly at about 545 nm and emits strongly at 572 nm instead and could be better suited for some instruments. B-Phycoerythrin may also be less "sticky" than R-Phycoerythrin and contributes less to background signal due to non-specific binding in certain applications.

Phycoerythrobilin

Phycoerythrobilin is a red phycobilin. Phycobilins are chromophores (light harvesting molecules) found in cyanobacteria and in the chloroplasts of red algae, glaucophytes and some cryptomonads. They are bonded to certain water-soluble proteins, known as phycobiliproteins.

Proteorhodopsin

Proteorhodopsin is a photoactive retinylidene protein in marine bacterioplanktons. Just like the homologous pigment bacteriorhodopsin found in some archaea, it consists of a transmembrane protein bound to a retinal molecule and functions as a light-driven proton pump. Some members of the family (of more than 800 types) are believed to have sensory functions.

Proteorhodopsin was first discovered in 2000. It was found in the genomes of several species of uncultivated marine γ -proteobacteria present in the Eastern Pacific Ocean, Central North Pacific Ocean and Southern Ocean, Antarctica. Subsequently, genes of proteorhodopsin variants have been identified in samples from the Mediterranean and Red Sea and the Sargasso Sea and the Sea of Japan.

On comparison to its better-known archaeal homolog bacteriorhodopsin, most of the active site residues of known importance to the bacteriorhodopsin mechanism are conserved in proteorhodopsin. Homologues of the active site residues Arg82, Asp85 (the primary proton acceptor), Asp212 and Lys216 (the retinal Schiff base binding site) in bacteriorhodopsin are conserved as Arg94, Asp97, Asp227 and Lys231 in proteorhodopsin. However, in proteorhodopsin, there are no carboxylic acid residues directly homologous to Glu194 or Glu204 of bacteriorhodopsin, which are thought to be involved in the proton release pathway at the extracellular surface.

It seems likely that proteorhodopsin functions throughout the earth's oceans as a light-driven H+ pump, by a mechanism similar to that of bacteriorhodopsin. As in bacteriorhodopsin, the retinal chromophore of bacteriorhodopsin is covalently bound to the apoprotein via a protonated Schiff base at Lys231. The configuration of the retinal chromophore in unphotolysed proteorhodopsin is predominantly all-trans, and changes to 13-cis upon illumination with light. Several models of the complete proteorhodopsin photocycle have been proposed, based on FTIR and UV–visible spectroscopy; they resemble established photocycle models for bacteriorhodopsin.

If the gene for proteorhodopsin is inserted into *E. coli* and retinal is given to these modified bacteria, then they will incorporate the pigment into their cell membrane and will pump protons in the presence of light.

Pigment Systems and Two Light Reactions

Electron transfers occur in pico-to-millisecond time range; they involve, in oxygenic photosynthesis, three major protein complexes: Photosystem II (water-plastoquinone oxido-reductase); Cytochrome b6/f complex (plastoquinol-plastocyanin oxido-reductase); and Photosystem I (plastocyanin-ferredoxin oxido-reductase). A fourth ATP synthase is required for ATP synthesis. An enigma, discovered by Emerson & Lewis, was that the quantum yield of oxygen evolution decline suddenly beyond 680 nm, when chlorophyll *a* was still absorbing light; this so-called "red-drop" phenomenon led to the current concept of the two light reactions and two pigment systems when Emerson *et al.* discovered the Enhancement effect in photosynthesis.

This concept is based on solid grounds: from the early ideas of E. Rabinowitch and of Hill & Bendall. In the experiments on the so-called Emerson Enhancement in photosynthesis, or in the Hill Reaction, yield in far-red light (light I) by simultaneous exposure to shorter wavelength light (light II), is enhanced. Further, the crucial experiments on the antagonistic effect of light I and II on the redox state of P700, on the Chlorophyll a fluorescence yield (Govindjee *et al.*, 1960; Butler, 1962; Duysens & Sweers, 1963), and, most importantly on the redox state of cytochrome f established the two light reaction scheme of photosynthesis. In fact, the existence of the series scheme of photosynthesis is best proven by light I (absorbed in photosystem I) oxidising cytochrome f and light II (absorbed in photosystem II) reducing the oxidised cytochrome f, However, without the fast absorption spectroscopic work on the new intermediate X-320 (now known to be equivalent to the so-called QA, the first bound plastoquinone molecule), the history of the discovery of the two light reactions would be incomplete.

Electron Transport: Acceptor Side of Photosystem II

A major advancement took place in the understanding of the functioning of the electron transfer pathway when Velthuys & Amesz (1974) and Bouges-Bocquet (1973) showed that there is a two electron gate on the acceptor side of PSII; a similar gate was discovered later in anoxygenic photosynthetic bacteria, independently by Vermeglio (1977) and by Wraight (1977). Electrons from the reaction centres are transferred first to a bound quinone molecule QA which is, surprisingly, a one-electron acceptor. The reduced QA transfers its electron to a loosely bound quinone QB; the reduced QB is tightly bound and has a long lifetime; it waits for a second electron until it can be doubly reduced to QB2-. The latter then becomes a quinol (QBH2) after grabbing two protons coming from the side of the membrane it is located on. The electron flow at the QAQB site can be written as:

QAQB+ light ⟶ QA–QB ⟶ QAQB-

QAQB- + light ⟶ QA QB2-

QA QB2-+ 2H+ ⟶ QA QBH2

QA QBH2 + PQ ⟶ QAQB + PQH2.

Although it is not certain, especially in plants, whether the protonation steps occur before the electron flow or *vice versa*, the above steps are referred to as the "two electron gate".

The Bicarbonate Effect

Over the past twenty-five years or more, we have been trying to understand the role of bicarbonate in the reactions on the two photosystems, a phenomenon discovered by Warburg & Krippahl (1960). Warburg (1964) took this phenomenon to suggest that oxygen in photosynthesis originated in CO_2, not in water, contradicting the existing idea that oxygen originated in water. He argued that (1) the 18_O experiments of Ruben *et al.* (1941) could not be trusted because of equilibration with both water and CO_2; (2) the hypothesis of C.B. van Niel was only from comparative biochemistry between oxygenic and anoxygenic organisms, and thus had no definitive value; and (3) the requirement of bicarbonate in the Hill reaction showed that it cannot be taken to prove that CO_2 was not the source of oxygen in photosynthesis. Bicarbonate seems to be involved not so much in the electron flow *per se*, but in the protonation steps. Interestingly, this effect is absent in anoxygenic photosynthetic bacteria. Differences between photosynthetic bacteria and photosystem II are, we believe, due to a preponderance of positively charged amino acids in the QB binding region of PS II over that in photosynthetic bacteria (Xiong *et al.*, 1996; 1998a). This may be responsible for different protonation channels in anoxygenic photosynthetic bacteria and in PS II. This bicarbonate site seems close to (Khanna *et al.*, 1981), but not identical to that where certain herbicides bind and act (Velthuys,1981; Wraight,1981). Terbutryn inhibits and kills plants by displacing QB, and bicarbonate seems to be involved in the protonation of the reduced form of QB.

One of the binding sites of bicarbonate has been shown to be on the non-haem iron and the other near an arginine residue, D1-R-257 (Xiong *et al.*, 1998b). It is not yet clear how this acceptor side binding would provide another major bicarbonate action on the oxygen evolving complex expounded by O. Warburg, H. Metzner and A. Stemler, and now being investigated in great detail by Klimov & co-workers (Allakhverdiev *et al.*, 1997; Hulsebosch *et al.*, 1998; Klimov *et al.*, 1997; Yruela *et al.*, 1998).

Oxygen Evolution

The answer to the question of the minimum number of quanta needed to evolve one oxygen molecule was solved in favour of 8-10 (Emerson & Lewis, 1943; Emerson, 1958; R. Govindjee *et al.*, 1968) rather than of 3 to 4 (Warburg & Negelein, 1923). This is in agreement with the scheme that two light reactions are needed for oxygen evolution and NADP+

reduction. A major breakthrough in the oxygen evolution steps was the discovery by Joliot *et al.* (1969) that oxygen/flash as a function of flash number oscillates with a period of 4 indicating accumulation of 4 positive charges on some intermediate before water is oxidised to molecular oxygen. A theory evolved by Kok *et al.*(1970) became the major framework, where the system started mostly in the S1 state in darkness, and oxygen release took place when S4 was converted to S0 (S representing the redox state of the tetranuclear Mn complex):

$$S1 + \text{light flash} \longrightarrow S2$$

$$S2 + \text{light flash} \longrightarrow S3$$

$$S3 + \text{light flash} \longrightarrow S4$$

$$S4 + 2H_2O \longrightarrow SO + O_2 + 4\,H^+$$

$$S0 + \text{light flash} \longrightarrow S1.$$

The elegant measurements of Dismukes & Siderer (1980) on multiline EPR (electron paramagnetic resonance) signals of Mn, and those of M. Klein and co-workers on EXAFS (Extended X-ray Absorption Fine Structure) and XAS (X-ray Absorption Spectroscopy) led to the identification of the S-state intermediates with mixed valence Mn complex (a tetranuclear Mn/PSII). The S0 is suggested to be Mn (II), Mn (III), Mn (IV), Mn (IV) (alternatively, Mn (III), Mn (III), Mn (III), Mn (IV)); S1 to be Mn (III), Mn (III), Mn (IV), Mn (IV); and both S2 & S3 as Mn (III), Mn (IV), Mn (IV), Mn (IV), Mn (IV). The mechanism of water oxidation to dioxygen is not yet known; however, there are theories as to how this occurs.

The recent finding of Messinger *et al.* (1997a) that S0 is also paramagnetic will further enhance our ability to understand the transitions of the S states, presented above. In addition, the possibility of showing that the S states can be reduced to S(-3) state lends credence to the current picture of the valence state of the Mn in the water oxidase. Coleman & Govindjee (1987) were the firstto suggest that the Mn atoms of the Mn cluster were bound to the reaction centre proteins D1 & D2; and, Kambara *et al.* (1985) and Padhye *et al.* (1986) were the first to suggest that an organic radical, specifically a histidine, may be the charge accumulator in addition to Mn during S2 to S3 transition.

The role of histidine is being actively investigated in several laboratories. Following the early pioneering work of W.F.J. Vermaas and of B. Diner and co-workers, Bowlby *et al.* (1996) have reported that Glutamic-65 and Histidine 337 on D1 protein may be ligands to Mn; and, Aspartic 103 and Glutamic 104, also on D1, may be ligands to Calcium. On the other hand, T. Wydrzynski and co-workers (Messinger *et al.*, 1995) are providing new information, using elegant mass spectroscopic methods, on the exchangeable water molecules at the site of water oxidation. Evidence for the proximity of a Calcium to the Mn cluster, has now been implied from X-ray absorption spectroscopy (Latimer *et al.*, 1995). Oxygen, the by-product of water oxidation, is released through the operation of the "oxygen clock" that involves 4 Mn atoms/PS II.

Thermoluminescence

A major tool to study the S-states was shown to be thermoluminescence, a phenomenon discovered by W. Arnold. Thermoluminescence (TL) is light emission, that originates mainly from PS II. If one shines light on a photosynthetic material, and freezes to 77K, all the separated charges are frozen, and heating the sample in darkness leads to thermoluminescence bands originating from the back reactions (i.e., recombination of charges) of say, e.g., S2 with QB- or S2 with QA-, etc. P.V. Sane, in collaboration with V.G. Tatake and T.S. Desai, was responsible for placing India on the World map of pioneering research in this area of thermoluminescence. The design of the instrument at the Bhabha Atomic Research Centre (BARC), Mumbai, India, where this work was done, provided very high Signal/Noise ratio. Several (6 or 7) TL bands were fully separated from each other and these bands, through separation of pigment systems I and II and through the ingenious use of inhibitors and electron carriers, were related to specific reactions in PSII and even PSI. It was because calculations, by Tatake *et al.* (1981), of the activation energies of the various TL bands did not match the Randall-Wilkins theory that led to the formulation by DeVault *et al.* (1983) of the most appropriate theory for TL in plants. Without the leadership provided by P.V. Sane, researches on TL at BARC would have been in directions different from those on the photosynthetic systems. Relationship of the S-states to the TL was, however, provided by Inoue &Shibata (1978) when they discovered the period 4 oscillation in TL and related it to oxygen evolution steps. Currently, the TL method is being used to monitor and characterise site-directed mutants both on the acceptor as well as the donor side of PS II.

Photosynthetic Antennas and Reaction Centres

Energy is first collected by an antenna system, and is transferred to a reaction centre complex where primary electron transfer takes place. Secondary reactions lead to oxidation of water and reduction of CO_2 in some classes of organisms. Antenna systems are highly regulated to maximise energy collection efficiency while avoiding photodamage.

Nature's most sophisticated and important solar energy storage system is found in photosynthetic organisms, including plants, algae and a variety of types of bacteria. All these organisms utilise sunlight to power cellular processes and ultimately derive most or all of their biomass through chemical reactions driven by light.

Photosynthesis begins when light is absorbed by an antenna pigment. This pigment can be a (bacterio)chlorophyll, carotenoid or bilin (open chain tetrapyrrole) depending on the type of organism. A wide variety of different antenna complexes are found in different photosynthetic systems. Antennas permit an organism to increase greatly the absorption cross-section for light without having to build an entire reaction centre and associated electron transfer system for each pigment, which would be very costly in terms of cellular resources. Energy transfer processes that may involve transfers to many intermediate pigments eventually results in the electronic excitation of a closely coupled pair of (bacterio)chlorophyll molecules in the photochemical reaction centre. The reaction centre is an integral membrane pigment-protein that carries out light-driven electron transfer reactions. The excited (bacterio)chlorophyll molecule transfers an electron to a nearby acceptor molecule, thereby creating an ion pair state consisting of the oxidised chlorophyll and reduced acceptor.

After the initial electron transfer event, a series of electron transfer reactions takes place that eventually stabilises the stored energy in forms that can be used by the cell. Some types of photosynthetic organisms have two different reaction centre complexes that work together in tandem, with the reduced acceptors of one photoreaction (photosystem 2) serving as the electron donor for the other centre (photosystem 1). In these organisms, the eventual electron donor is water, liberating molecular oxygen, and the ultimate electron acceptor is carbon dioxide, which is reduced to sugars. Other types of photosynthetic organisms contain only a single photosystem, which in some cases is more similar to photosystem 2 and in other cases to photosystem 1 of the oxygen-evolving organisms. Oxygen is not produced in any of the naturally occurring single photosystem organisms, which are therefore called anoxygenic.

Reaction Centre: Structure and Function

The reaction centre complexes from the anoxygenic purple photosynthetic bacteria are the best understood of all photosynthetic reaction centres, from both a structural and a functional point of view. These were the first reaction centre complexes to be purified, the first to be studied by picosecond kinetic methods and the first to have X-ray structures solved. Much of the molecular level understanding of the early events in photosynthesis is based on the information derived from these systems.

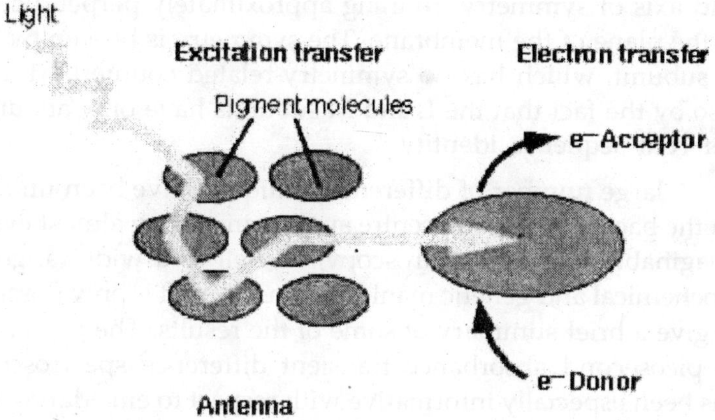

Light

Excitation transfer

Electron transfer

Pigment molecules

e⁻Acceptor

e⁻Donor

Antenna

Figure: Basic concept of photosynthetic antenna and reaction centre function

The reaction centres from purple photosynthetic bacteria contain a core protein complex consisting of two related yet distinct integral membrane proteins, known as L (Light) and M (Medium). Most also contain a third protein, known as H (Heavy), and some contain a fourth subunit known as C (cytochrome). The C subunit is a four-haem containing *c*-type cytochrome.

In addition to the protein complement, these reaction centres contain several additional co-factors, that are not covalently attached to the protein. These include bacteriochlorophyll *a* (in some cases *b*), the corresponding metal-free bacteriopheophytins, two quinones (either ubiquinone or menaquinone), a non-haem Fe, and in most cases a molecule of carotenoid. The cytochrome subunit contains four haem *c* groups, covalently bound to cysteine residues.

The reaction centre protein forms a scaffolding upon which the co-factors are arranged. The part of the protein that crosses the lipid bilayer is almost purely alpha helical in secondary structure, and contains predominantly non-polar amino acids, with almost no charged amino acids. There are 11 transmembrane helices, with 5 each from L and M, and one from the H subunit. The L and M proteins have a pseudo-2 fold axis of symmetry, running approximately perpendicular to the plane of the membrane. The symmetry is broken by the H subunit, which has no symmetry-related counterpart, and also by the fact that the L and M subunits have only about 60 per cent sequence identity.

A large number of different techniques have been utilised on the bacterial reaction centre system, including almost every imaginable kind of spectroscopy, as well as a wide range of biochemical and genetic manipulations. Here it is only possible to give a brief summary of some of the results. The technique of picosecond absorbance transient difference spectroscopy has been especially informative with respect to elucidating the pathway of electron flow in these complexes. Figure summarises the photochemical and early secondary reactions that take place in isolated reaction centres. A variety of evidence indicates that the electron transfer pathway and kinetics in isolated reaction centres are not significantly altered from their behaviour in vivo.

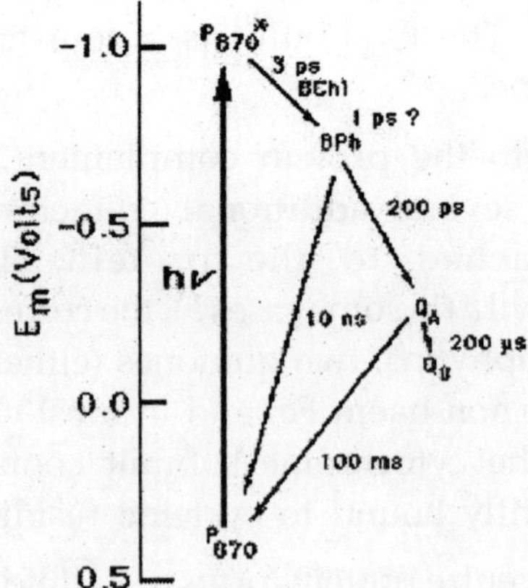

Figure: Energy-kinetic diagram for reaction centres from purple photosynthetic bacteria

Perhaps the most striking aspect of the comparison of the structural and kinetic data on reaction centres is the apparent two-fold symmetry of the structure, contrasted with the clear evidence for asymmetry in the electron transfer pathway.

Another interesting feature of the rate constants shown in Figure is that the rates of recombination processes, in which the electron simply returns from one of the acceptors directly to the oxidised special pair, are invariably a factor of fifty or more slower than are the rate constants for the forward reaction. It is because of this kinetic control of the rates of wasteful processes that the quantum yield for photochemistry is so high. Again, the molecular reasons for this remarkable behaviour have proven difficult to elucidate. One of the major factors thought to be involved in the slow recombination rates is the large energy gap involved, placing these reactions in the Marcus inverted region of the rate vs free energy curve.

Other types of photosynthetic bacteria contain reaction centre complexes that are structurally quite different from those of the purple bacteria, and our understanding of them is at a much more primitive state. Similarly, structural information on the photosystems from oxygenic organisms is less detailed than that from purple bacteria. It is clear that all photosynthetic reaction centres operate on similar overall principles, so the general lessons learned from the purple bacteria are almost certainly largely applicable to other systems. However, some systems include activities that are not found in the purple bacteria, such as oxygen evolution by photosystem 2. Therefore, study of more than one type of organism is necessary to understand the wide scope of processes that are carried out by photosynthetic reaction centres.

Function and Regulation of Antenna Systems

The vast majority of the pigments in a photosynthetic organism are not chemically active, but function primarily as an antenna. The photosynthetic antenna system is organised to collect and deliver excited state energy by means of excitation transfer to the reaction centre complexes where photochemistry takes place. The antenna system increases the effective cross-section of photon absorption by increasing the number of pigments associated with each photochemical complex. The intensity of sunlight is sufficiently dilute so that any given chlorophyll molecule only absorbs at most a few photons per second. By incorporating many pigments into a single unit, the biosynthetically expensive reaction centre and electron transport chain can be used to maximum efficiency. A remarkable variety of antenna complexes have been identified from various classes of photosynthetic organisms. There seems to be little doubt that there have been multiple evolutionary origins of antenna complexes, as there is no common structural theme evident.

Excitation transfer must be fast enough to deliver excitations to the photochemical reaction centre and have them trapped in a time short compared to the excited state lifetime in the absence of trapping. Excited state lifetimes of isolated antenna complexes, where the reaction centres have been removed, are typically in the 1-5 ns range. Observed excited state lifetimes of systems where antennas are connected to reaction centres are generally on the order of a few tens of picoseconds, which is sufficiently fast so that under physiological conditions almost all the energy is trapped by photochemistry.

Antenna systems are often viewed as being "on" all the time, with the regulation of photosynthesis in response to different conditions taking place primarily in the reaction centres and carbon metabolism enzymes. Clearly, this is not the case, and the modern view is of a much more actively regulated system at all stages of energy storage. The advantages of "directional signals" or "volume controls" to regulate either the distribution between the photosystems or the number of excitations delivered by the entire antenna network are easy to appreciate.

One of the most interesting and important of these regulatory mechanisms is the phenomenon of "non-photochemical quenching" (qN) of chlorophyll excited states in chloroplasts.

During periods of high irradiance such as mid-day, or under certain stress conditions, a substantial fraction of the excited state energy is dissipated by quenching before it is ever transferred to the reaction centre.

The basic idea is that it is much easier and safer for cells to dispose of this energy before it initiates the photochemical processes in reaction centres than it is for them to try to repair the substantial photo-oxidative damage that can result from excess light. This process is now thought to be a major regulatory mechanism and understanding, it is likely to have great economic significance.

Considerable evidence indicates that a cycle involving carotenoids known as xanthophylls plays a role in this regulation, with zeaxanthin associated with the quenched state and violaxanthin associated with the non-quenched state. Enzymes interconvert these carotenoids in response to energetic signals generated in the chloroplast.

The chemical mechanism of the quenching effect is not yet understood in molecular detail. One proposal under investigation is that the zeaxanthin molecule with its more extensive conjugation has a lower excited state energy than does violaxanthin.

The excited state energies are proposed to be such that violaxanthin lies above chlorophyll and therefore acts as an energy donor while zeaxanthin lies below chlorophyll and therefore acts as an excited state quencher. Other factors are also known to be important, including the state of aggregation of the antenna complex and the pH of the interior lumen region of the thylakoid membrane.

The chlorophyll *a/b*-containing antenna complex known as LHC II is thought to be the site of much of this quenching reaction. This complex is well characterised structurally. While the linkage of the quenched state to the presence of zeaxanthin in the membrane is clear, it is mostly based on evidence showing that the two are correlated, rather than an unambiguous cause and effect relationship.

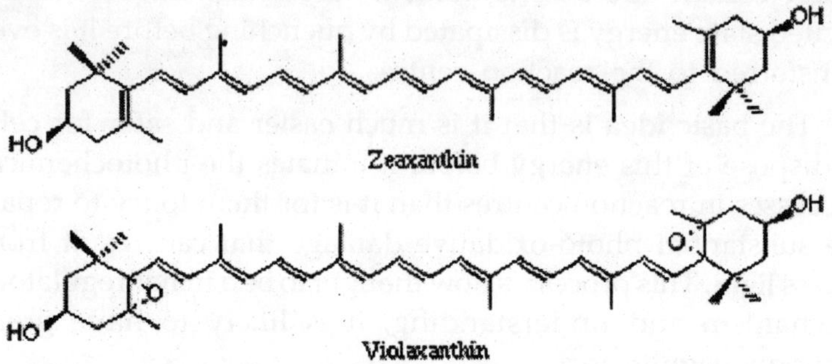

Zeaxanthin

Violaxanthin

Figure: Molecular structures of zeaxanthin and violaxanthin

Future Challenges

There are many fundamental unsolved questions relating to the mechanism of energy storage by natural photosynthesis. Certain of these questions directly relate to the effort to develop artificial photochemically based systems for solar energy storage, in which water is oxidised to molecular oxygen, along with either hydrogen production or CO_2 reduction. Chief among these is developing a deep understanding of the physical principles by which all natural photosynthetic systems avoid the wasteful recombination reactions involving early radical ion pair intermediates. A second important area includes the elucidation of the details of the reaction centre structure and chemical mechanism of water oxidation by photosystem 2. A third area involves structure and function of photosystem 1, including how the strongly reducing early electron acceptors avoid auto-oxidation by oxygen.

Finally, a broad area that is presently very poorly understood includes the dynamic regulation of natural antenna systems so as to maximise energy collection efficiency under a variety of conditions while at the same time avoiding harmful overexcitation and subsequent damage. Ultimately, an understanding of these principles that underlie natural photosynthesis should point the way to the development of efficient, robust artificial systems for solar energy storage.

Light Reactions

Light-dependent Reaction

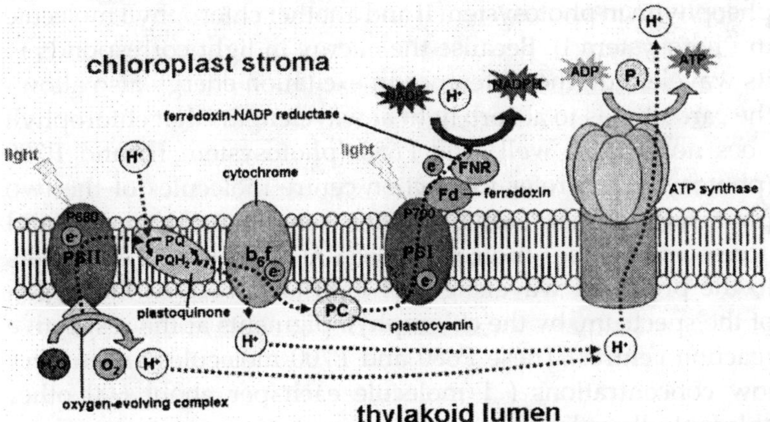

Figure: Light-dependent reactions of photosynthesis at the thylakoid membrane

The first stage of the photosynthetic system is the *light-dependent reaction,* which converts solar energy into chemical energy.

The light dependent reaction produces oxygen gas and converts ADP and NADP+ into the energy carriers ATP and NADPH.

The process of synthesising ATP and NADPH is accomplished via the mechanism of an electron transport chain. This is a series of proteins embedded in a biological membrane that transfers high-energy electrons from one to another, accomplishing various activities along the way as the electron drops in energy level. When sunlight strikes a cluster of chlorophyll molecules, the molecule is dulled, and an electron is switched to a higher energy level of the molecule. The excitation is transferred as an exciton from one antenna chlorophyll to another until it is captured by a primary reaction centre. It can be transferred from one molecule to another of the same kind of pigment, or from a carotenoid to chlorophyll, but not from chlorophyll to a carotenoid, because excitation of carotenoids carries more energy than that of chlorophyll. Only chlorophyll of the reaction centre is capable of transferring an electron to an electron acceptor (an intermediate, e.g., pheophytin in photosystem II and another chlorophyll molecule in Photosystem I). Because the energy in light corresponds to its wavelength, the difference in excitation energy also allows the carotenoids to absorb light at wavelengths that chlorophyll does not absorb well. The P680 (photosystem II) and P700 (photosystem I) refer to reaction centre molecules of the two photosystems that receive excitons from other chlorophyll and accessory pigment molecules; each number 680 or 700 refers to the preferred wavelength of light absorbed, in the region of the spectrum, by the chlorophyll pigments at the respective reaction centres. These P680 and P700 molecules are in very low concentrations (1 molecule each per about 600 other chlorophyll molecules).

The rate of this stage of the light-dependent reactions can be monitored with the dye DPIP, or ferricyanide or methyl viologen, which accepts some of the electrons that would normally go to NADPH and changes colour as a result.

The chlorophyll's electron can follow either of two different pathways, cyclic or non-cyclic.

Cyclic Photophosphorylation

In cyclic electron flow, the electron begins in a pigment complex called photosystem I, passes from the primary acceptor to ferredoxin, then to a complex of two cytochromes (similar to those found in mitochondria), and then to plastocyanin before returning to chlorophyll. This transport chain produces a proton-motive force, pumping H^+ ions across the membrane; this produces a concentration gradient which can be used to power ATP synthase during chemiosmosis. This pathway is known as cyclic photophosphorylation, and it produces neither O_2 nor NADPH. In bacterial photosynthesis, a single photosystem is used, and therefore is involved in cyclic photophosphorylation.

Non-cyclic Photophosphorylation

The other pathway, non-cyclic photophosphorylation, is a two-stage process involving two different chlorophyll photosystems. First, a water molecule is broken down into $2H^+ + 1/2O_2 + 2e^-$. The two electrons from the water molecule are kept in photosystem II, while the $2H^+$ and $1/2O_2$ are left out for further use. Then a photon is absorbed by the chlorophyll core of photosystem II, exciting the two electrons which are transferred to the acceptor molecule. The deficit of electrons is replenished by taking electrons from another molecule of water. The electrons transfer from the primary acceptor to plastoquinone, then to plastocyanin, producing proton-motive force as with cyclic electron flow and driving ATP synthesis.

The photosystem II complex replaced its lost electrons from an external source, however, the two other electrons are not returned to photosystem II as they would in the analogous cyclic pathway. Instead, the still-excited electrons are transferred to a photosystem I complex, which boosts their energy level to a higher level using a second solar photon. The highly excited electrons are transferred to the acceptor molecule, but this time are passed on to an enzyme called Ferredoxin-NADP reductase/$NADP^+$ reductase, for short FNR, which uses them to catalyst the reaction (as shown):

$$NADP^+ + 2H^+ + 2e^- \rightarrow NADPH + H^+$$

This consumes the H^+ ions produced by the splitting of water, leading to a net production of $1/2O_2$, ATP, and $NADPH+H^+$ with the consumption of solar photons and water.

The concentration of NADPH in the chloroplast may help regulate, which pathway electrons take through the light reactions. When the chloroplast runs low on ATP for the Calvin cycle, NADPH will accumulate and the plant may shift from non-cyclic to cyclic electron flow.

Steps: It is important to note that both photosystems are almost simultaneously excited; thus, both photosystems begin functioning at almost the same time.

1. Light strikes photosystem II and the energy is absorbed and passed along until it reaches P680 chlorophyll.
2. The excited electron is passed to the primary electron acceptor. Photolysis in the thylakoid takes the electrons from water and replaces the P680 electrons that were passed to the primary electron acceptor. (O_2 is released as a waste product)
3. The electrons are passed to photosystem I via the electron transport chain (ETC) and in the process used to pump protons across the thylakoid membrane into the lumen.
4. The stored energy in the proton gradient is used to produce ATP which is used later in the Calvin-Benson Cycle.
5. P700 chlorophyll then uses light to excite the electron to its second primary acceptor.
6. The electron is sent down another ETC and used to reduce $NADP^+$ to NADPH.
7. The NADPH is then used later in the Calvin-Benson Cycle.

Oxygen Evolving Complex

The oxygen evolving complex is a water oxidising enzyme involved in the photo-oxidation of water during the light reactions of photosynthesis. Based on a widely accepted theory from 1970, the complex can exist in 5 states: S_0 to S_4. Photons trapped by photosystem II move the system from state S_0 to S_4. S_4 is unstable and reacts with water to produce oxygen. Currently the functional mechanism of the complex is not completely understood. Much of the known data have been collected from flash experiments, EPR, and X-ray spectroscopy.

P680: P680 is the reaction-centre chlorophyll *a* molecule associated with photosystem II. Its absorption spectrum peaks at 680 nm, which is in the red part of the spectrum. When photosystem II absorbs light, an electron is excited to a higher energy level in the P680 chlorophyll. The resulting P680 with excited electron is designated P680*. The electron is subsequently captured by the primary electron acceptor.

P680 is the strongest biological oxidising agent known.

Photophosphorylation

The production of ATP using the energy of sunlight is called photophosphorylation. Only two sources of energy are available to living organisms: sunlight and oxidation-reduction (redox) reactions. All organisms produce ATP, which is the universal energy currency of life.

In photophosphorylation, light energy is used to create a high-energy electron donor and a lower-energy electron acceptor. Electrons then move spontaneously from donor to acceptor through an electron transport chain.

Background: ATP is made by an enzyme called ATP synthase. The structure of this enzyme and its underlying genetic code is remarkably similar in all known forms of life.

ATP synthase is powered by a transmembrane electrochemical potential gradient, usually in the form of a proton gradient. The function of the electron transport chain is to produce this gradient. In all living organisms, a series of redox reactions is used to produce a transmembrane electrochemical potential gradient, or a so called proton motive force (pmf).

Redox reactions are chemical reactions in which electrons are transferred from a donor molecule to an acceptor molecule. The underlying force driving these reactions is the Gibbs free energy of the reactants and products. The Gibbs free energy is the energy available "free" to do work. Any reaction that decreases the overall Gibbs free energy of a system will proceed spontaneously. The transfer of electrons from a high-energy molecule (the donor) to a lower-energy molecule (the acceptor) can be *spatially* separated into a series of intermediate redox reactions. This is an electron transport chain.

The fact that a reaction is thermodynamically possible does not mean that it will actually occur. A mixture of hydrogen gas and oxygen gas does not spontaneously ignite. It is necessary either to supply an activation energy, or to lower the intrinsic activation energy of the system, in order to make most biochemical reactions proceed at a useful rate. Living systems use complex macromolecular structures (enzymes) to lower the activation energies of biochemical reactions.

It is possible to couple a thermodynamically favourable reaction (a transition from a high-energy state to a lower-energy state) to a thermodynamically unfavourable reaction (such as a separation of charges, or the creation of an osmotic gradient), in such a way that the overall free energy of the system decreases (making it thermodynamically possible), while useful work is done at the same time. Biological macromolecules that catalyse a thermodynamically favourable reaction *if and only if a thermodynamically unfavourable reaction occurs simultaneously* underlie all known forms of life.

Electron transport chains produce energy in the form of a transmembrane electrochemical potential gradient. This energy is used to do useful work. The gradient can be used to transport molecules across membranes. It can be used to do mechanical work, such as rotating bacterial flagella. It can be used to produce ATP and NADPH, high-energy molecules that are necessary for growth.

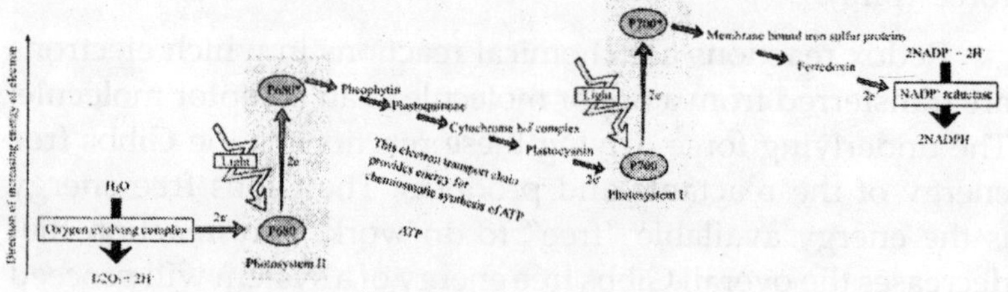

The cells of all plants and photosynthetic algae contain chloroplasts, which produce ATP and NADPH using the energy of sunlight. Chloroplasts have a deep resemblance to certain prokaryotes, which suggests that chloroplasts originated as intracellular symbionts in primitive eukaryotic cells.

The electron transport chain in chloroplasts is contained in two extremely complex transmembrane structures, *Photosystem II* (PS II) and *Photosystem I* (PS I). PS II and PS I are linked by a transmembrane proton pump, cytochrome b_6f, which is similar to mitochondrial *Complex III*.

The overall process is the transfer of electrons from water to NADPH via a transmembrane proton pump:

$$H_2O \rightarrow \text{PS II} \rightarrow \text{plastoquinone} \rightarrow b_6f \rightarrow \text{plastocyanin} \rightarrow \text{PS I} \rightarrow \text{NADPH}$$

Photosystem II

PS II is an extremely complex, highly organised transmembrane structure that contains a *water-splitting complex*, chlorophylls *a* and *b*, a *reaction centre* (P680), pheophytin (a pigment similar to chlorophyll), and two quinones. It uses the energy of sunlight to transfer electrons from water to a mobile electron carrier in the membrane called *plastoquinone*:

$$H_2O \rightarrow P680 \rightarrow P680^* \rightarrow \text{plastoquinone}$$

Plastoquinone, in turn, transfers electrons to b_6f, which feeds them into PS I.

Water-splitting Complex

The step $H_2O \rightarrow P680$ is performed by a poorly-understood structure embedded within PS II called the *water-splitting complex* or the *oxygen-evolving complex*. It catalyses a reaction that splits water into electrons, protons and oxygen:

$$2H_2O \rightarrow 4H^+ + 4e^- + O_2$$

The electrons are transferred to special chlorophyll

molecules (embedded in PS II) that are promoted to a higher-energy state by the energy of photons.

Light-harvesting System

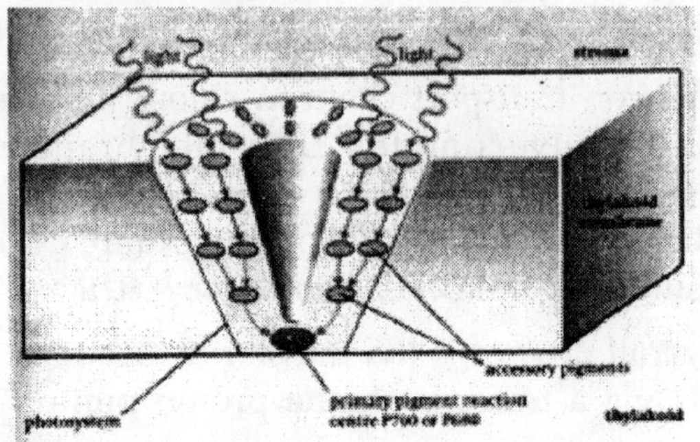

A Photosystem: A light-harvesting cluster of photosynthetic pigments present in the thylakoid membrane of chloroplasts.

The light harvesting-system is composed of antenna proteins or antenna pigments which are complex structures embedded in the membrane that absorb energy from sunlight and transfer it to special chlorophyll molecules in PS II. There are hundreds of antenna proteins for every PS II. The transfer process (*exciton transfer*) is extremely efficient, due to the ability of antenna proteins to transfer their excitation energy to neighbouring molecules in a quantum, all-or-none fashion. Most of the energy gathered by antenna proteins is ultimately transferred to special reaction centre proteins P680 in PS II, which are promoted to a high-energy state.

Reaction Centre

The excitation P680 $\rightarrow$ P680* of the reaction centre pigment P680 occurs here. These special chlorophyll molecules embedded in PS II absorb the energy of photons, with maximal absorption at 680 nm. Electrons within these molecules are promoted to a higher-energy state. This is one of two core processes in photosynthesis, and it occurs with astonishing efficiency (greater than 90 per cent) because, in addition to direct excitation by light at 680 nm, the energy of light first harvested by *antenna proteins* at other wavelengths in the light-harvesting system is also transferred to these special chlorophyll molecules.

This is followed by the step P680* → plastoquinone which occurs within the reaction centre of PS II. High-energy electrons are transferred to plastoquinone. Plastoquinone is then released into the membrane as a mobile electron carrier.

This is the second core process in photosynthesis. The initial stages occur within *picoseconds*, with an efficiency of 100 per cent. The seemingly impossible efficiency is due to the precise positioning of molecules within the reaction centre. This is a solid-state process, not a chemical reaction. It occurs within an essentially crystalline environment created by the macromolecular structure of PS II. The usual rules of chemistry (which involve random collisions and random energy distributions) do not apply in solid-state environments.

PS II is a transmembrane structure found in all chloroplasts. It splits water into electrons, protons and molecular oxygen. The electrons are transferred to plastoquinone, which carries them to a proton pump. Molecular oxygen is released into the atmosphere.

The emergence of such an incredibly complex structure, a macromolecule that converts the energy of sunlight into potentially useful work with efficiencies that are impossible in ordinary experience, seems almost magical at first glance. Thus it is of considerable interest that essentially the same structure is found in *purple bacteria*.

Cytochrome b_6f

PS II and PS I are connected by a transmembrane proton pump, cytochrome b_6f complex (plastoquinol—plastocyanin reductase; EC 1.10.99). Electrons from PS II are carried by plastoquinone to b_6f, where they are removed in a stepwise fashion and transferred to a water-soluble electron carrier called *plastocyanin*. This redox process is coupled to the pumping of four protons across the membrane. The resulting proton gradient (together with the proton gradient produced by the water-splitting complex in PS II) is used to make ATP via ATP synthase.

The similarity in structure and function between cytochrome b_6f (in chloroplasts) and cytochrome bc_1 (*Complex III* in mitochondria) is striking. Both are transmembrane structures that remove electrons from a mobile, lipid-soluble electron carrier (plastoquinone in chloroplasts; ubiquinone in mitochondria) and transfer them to a mobile, water-soluble electron carrier (plastocyanin in chloroplasts; cytochrome *c* in mitochondria). Both are proton pumps that produce a transmembrane proton gradient.

Photosystem I

PS I accepts electrons from plastocyanin and transfers them either to NADPH (*noncyclic electron transport*) or back to cytochrome b_6f (*cyclic electron transport*):

plastocyanin → P700 → P700* → ferredoxin → NADPH

↑ ↓

b_6f ← plastoquinone b_6f

PS I, like PS II, is a complex, highly organised transmembrane structure that contains antenna chlorophylls, a reaction centre (P700), phylloquinine, and a number of iron-sulphur proteins that serve as intermediate redox carriers.

The light-harvesting system of PS I uses multiple copies of the same transmembrane proteins used by PS II. The energy of absorbed light (in the form of delocalised, high-energy electrons) is funnelled into the reaction centre, where it excites special chlorophyll molecules (P700, maximum light absorption at 700 nm) to a higher energy level. The process occurs with astonishingly high efficiency.

Electrons are removed from excited chlorophyll molecules and transferred through a series of intermediate carriers to *ferredoxin*, a water-soluble electron carrier. As in PS II, this is a solid-state process that operates with essentially 100 per cent efficiency.

There are two different pathways of electron transport in PS I. In *non-cyclic electron transport*, ferredoxin carries the electron to the enzyme ferredoxin $NADP^+$ oxidoreductase that reduces $NADP^+$ to NADPH. Alternately, in *cyclic electron transport*, electrons from ferredoxin are transferred (via plastoquinone) to a proton pump, cytochrome b_6f. They are then returned (via plastocyanin) to P700.

NADPH and ATP are used to synthesise organic molecules from CO_2. The ratio of NADPH to ATP production can be adjusted by adjusting the balance between cyclic and non-cyclic electron transport.

It is noteworthy that PS I closely resembles photosynthetic structures found in *green sulphur bacteria*, just as PS II resembles structures found in *purple bacteria*.

Photosynthetic Electron Transport Chains in Bacteria

PS II, PS I and cytochrome b_6f are found in chloroplasts. All plants and all photosynthetic algae contain chloroplasts, which produce NADPH and ATP by the mechanisms described above. Essentially the same transmembrane structures are also found in *cyanobacteria*.

Unlike plants and algae, cyanobacteria are prokaryotes. They do not contain chloroplasts. Rather, they bear a striking resemblance to chloroplasts themselves. This suggests that organisms resembling cyanobacteria were the evolutionary precursors of chloroplasts. One imagines primitive eukaryotic cells taking up cyanobacteria as intracellular symbionts.

Cyanobacteria

Cyanobacteria contain structures similar to PS II and PS I in chloroplasts. Their light-harvesting system is different from that found in plants (they use *phycobilins*, rather than chlorophylls, as antenna pigments), but their electron transport chain is essentially the same as the electron transport chain in chloroplasts. The mobile water-soluble electron carrier is cytochrome c_6 in cyanobacteria, plastocyanin in plants.

Cyanobacteria can also synthesise ATP by oxidative phosphorylation, in the manner of other bacteria. The electron transport chain is:

NADH dehydrogenase → plastoquinone →
b_6f → cytochrome c_6 → cytochrome aa_3 → O_2

where the mobile electron carriers are plastoquinone and cytochrome c_6, while the proton pumps are NADH dehydrogenase, b_6f and cytochrome aa_3.

Cyanobacteria are the only bacteria that produce oxygen during photosynthesis. The Earth's primordial atmosphere was anoxic. Organisms like cyanobacteria produced our present-day oxygen containing atmosphere.

The other two major groups of photosynthetic bacteria, purple bacteria and green sulphur bacteria, contain only a single photosystem and do not produce oxygen.

Purple Bacteria

Purple bacteria contain a single photosystem that is structurally related to PS II in cyanobacteria and chloroplasts:

P870 → P870* → ubiquinone → bc_1 →
cytochrome c_2 → P870

This is a *cyclic* process in which electrons are removed from an excited chlorophyll molecule (*bacteriochlorophyll*; P870), passed through an electron transport chain to a proton pump (cytochrome bc_1 complex, similar but not identical to cytochrome bc_1 in chloroplasts), and then returned to the cholorophyll molecule. The result is a proton gradient, which is used to make ATP via ATP synthase. As in cyanobacteria and chloroplasts, this is a solid-state process that depends on the precise orientation of various functional groups within a complex transmembrane macromolecular structure.

In order to make NADPH, purple bacteria use an external electron donor (hydrogen, hydrogen sulphide, sulphur, sulphite, or organic molecules such as succinate and lactate) to feed electrons into a reverse electron transport chain.

Green Sulphur Bacteria

Green sulphur bacteria contain a photosystem that is analogous to PS I in chloroplasts:

$$\begin{array}{ccccccc} P840 & \rightarrow & P840^{*} & \rightarrow & \text{ferredoxin} & \rightarrow & \text{NADH 1} \\ \uparrow & & \downarrow & & & & \\ \text{cyt } c_{553} & \leftarrow & bc_1 & \leftarrow & \text{menaquinone} & & \end{array}$$

There are two pathways of electron transfer. In *cyclic electron transfer*, electrons are removed from an excited chlorophyll molecule, passed through an electron transport chain to a proton pump, and then returned to the chlorophyll. The mobile electron carriers are, as usual, a lipid-soluble quinone and a water-soluble cytochrome. The resulting proton gradient is used to make ATP.

In *non-cyclic electron transfer*, electrons are removed from an excited chlorophyll molecule and used to reduce NAD^+ to NADH. The electrons removed from P840 must be replaced. This is accomplished by removing electrons from H_2S, which is oxidised to sulphur (hence the name "green sulphur bacteria"). Purple bacteria and green sulphur bacteria occupy relatively minor ecological niches in the present day biosphere. They are of interest because of their importance in pre-cambrian ecologies, and because they were the evolutionary precursors of modern plants.

Photosynthetic Reaction Centre

A photosynthetic reaction centre is a protein that is the site of the light reactions of photosynthesis. The reaction centre contains pigments such as chlorophyll and phaeophytin. These absorb light, promoting an electron to a higher energy level within the pigment. The free energy created is used to reduce an electron acceptor, and is critical for the production of chemical energy during photosynthesis. Reaction centres are present in all green plants and in many bacteria and algae. Green plants have two reaction centres known as photosystem I and photosystem II and the structures of these centres are complex, involving a multi-subunit protein. The reaction centre found in *Rhodopseudomonas* bacteria is currently better understood since it has fewer proteins than the examples in green plants.

Capturing Light Energy

A reaction centre is laid out in such a way that it captures the energy of a photon using pigment molecules and turns it into a usable form. Once the light energy has been absorbed directly by the pigment molecules, or passed to them by resonance transfer from antenna pigments, they release two electrons into an electron transport chain.

Light is made up of small bundles of energy called photons. If a photon with the right amount of energy hits an electron it will raise the electron to a higher energy level. Electrons are most stable at their lowest energy level or ground state, the orbit in which the electron has the least amount of energy. Electrons in higher energy levels can return to ground state in a manner analogous to a ball falling down a staircase. In doing so they release energy. This is the process which is exploited by a photosynthetic reaction centre.

When an electron rises to a higher energy level it increases the reduction potential of the molecule. This means it has a greater tendency to donate electrons, the key to the conversion of light energy to chemical energy. In green plants, the electron transport chain that follows has many electron acceptors including phaeophytin, quinone, plastoquinone, cytochrome bf, and ferredoxin that ultimately result in the reduced molecule NADPH. The passage of the electron through the electron transport chain also results in the pumping of protons (hydrogen ions) from the chlorplast's stroma into the lumen resulting in a proton gradient across the thylakoid membrane that can be used to synthesis ATP using ATP synthase. Both the ATP and NADPH are used in the Calvin cycle to fix carbon dioxide into triose sugars.

Bacteria

The bacterial photosynthetic reaction centre has been an important model to understand the structure and chemistry of the biological process of capturing light energy. In the 1960s, Roderick Clayton was the first to purify the reaction centre complex from purple bacteria. However, the first crystal structure was determined by Hartmut Michel, Johann Deisenhofer and Robert Huber for which they shared the Nobel Prize in 1988. This was also significant since it was the first structure for any membrane protein complex.

The reaction centre contains two pigments that serve to collect and transfer the energy from photon absorption: BChb and Bph. BChb roughly resembles the chlorophyll molecule found in green plants, but due to minor structural differences, its peak absorption wavelength is shifted into the infra-red, with wavelengths as long as 1000nm. Bph has the same structure as BChb, but the central magnesium ion is replaced by two protons.

Mechanism

The process starts when light is absorbed by two BChl-b molecules that lie near the periplasmic side of the membrane. This pair of chlorophyll molecules, often called the "special pair", absorbs photons at roughly 960nm, and thus is called P960 (with *P* standing for "pigment"). Once P960 absorbs a photon it ejects an electron, which is transferred through another molecule of Bchl to the BPh in the L subunit. This initial charge separation yields a positive charge on P960 and a negative charge on the BPh. This process takes place in 10 picoseconds (10^{-11} seconds).

The charges on the $P960^+$ and the BPh^- could undergo charge recombination in this state. This would waste the high-energy electron and convert the absorbed light energy into heat. Several factors of the reaction centre structure serve to prevent this. First the transfer of an electron from BPh^- to $P960^+$ is relatively slow compared to two other redox reactions in the reaction centre. The faster reactions involve the transfer of an electron from BPh^- (BPh^- is oxidised to BPh) to the electron acceptor quinone (Q_A) and the transfer of an electron to $P960^+$ ($P960^+$ is reduced to P960) from a haem in the cytochrome subunit above the reaction centre.

The high-energy electron which resides on the tightly bound quinone molecule Q_A is transferred to an exchangeable quinone molecule Q_B. This molecule is loosely associated with the protein and is fairly easy to detach. Two of the high-energy electrons are required to fully reduce Q_B to QH_2 taking up two protons from the cytoplasm in the process. The reduced quinone QH_2 diffuses through the membrane to another protein complex (cytochrome bc_1-complex) where it is oxidised. In the process the reducing power of the QH_2 is used to pump protons across the membrane to the periplasmic space. The electrons from the cytochrome bc_1-complex are then transferred through a soluble cytochrome c intermediate, called cytochrome c_2, in the periplasm to the cytochrome subunit. Thus, the flow of electrons in this system is cyclical.

Green Plants

Oxygenic Photosynthesis

In 1772, chemist Joseph Priestly carried out a series of experiments relating to the gasses involved in respiration and combustion. In his first experiment, he lit a candle and placed it under an upturned jar. After a short period of time, the candle burned out. He carried out a similar experiment with a mouse in the confined space of the burning candle. He found that the mouse died a short time after the candle had been extinguished. However, he could revivify the foul air by placing green plants in the area and exposing them to light. Priestly's observations were some of the first experiments that demonstrated the activity of a photosynthetic reaction centre.

In 1779, Jan Ingenhousz carried out more than 500 experiments spread out over 4 months in an attempt to understand what was really going on. He wrote up his discoveries in a book entitled 'Experiments upon Vegetables'. Ingenhousz took green plants and immersed them in water inside a transparent tank. He observed many bubbles rising from the surface of the leaves whenever the plants were exposed to light. Ingenhousz collected the gas which was given off by the plants and performed several different tests in attempt to determine what the gas was. The test which finally revealed the identity of the gas was placing a smoldering taper into the gas sample and having it relight. This test proved it was oxygen, or as Joseph Priestly had called it, 'de-phlogisticated air'.

In 1932, Professor Robert Emerson and an undergraduate student, William Arnold, used a repetitive flash technique to precisely measure small quantities of oxygen evolved by chlorophyll in the algae *Chlorella*. Their experiment proved the existence of a photosynthetic unit. Gaffron and Wohl later interpreted the experiment and realised that the light absorbed by the photosynthetic unit was transferred. This reaction occurs at the reaction centre of photosystem II and takes place in cyanobacteria, algae and green plants.

Photosystem II

Photosystem II is the photosystem that generates the electron that will eventually reduce $NADP^+$. Photosystem II is present on the thylakoid membranes inside chloroplasts, the site of photosynthesis in green plants. The structure of Photosystem II is remarkably similar to the bacterial reaction centre and it is theorised that they share a common ancestor.

The core of photosystem II consists of two subunits referred to as D1 and D2. These two subunits are similar to the L and M subunits present in the bacterial reaction centre. Photosystem

Photosystem II differs from the bacterial reaction centre in that it has many additional subunits which bind additional chlorophylls to increase efficiency. The overall reaction catalysed by photosystem II is:

$$2Q + 2H_2O \overset{light}{\Rightarrow} O_2 + 2QH_2$$

Q represents plastoquinone, the oxidised form of Q. QH_2 represents plastoquinol, the reduced form of Q. This process of reducing quinone is comparable to that which takes place in the bacterial reaction centre. Photosystem II obtains electrons by oxidising water in a process called photolysis. Molecular oxygen is a by-product of this process and it is this reaction that supplies the atmosphere with oxygen. The fact that the oxygen from green plants originated from water was first deduced by the Canadian-born American biochemist Martin David Kamen. He used a radioactive isotope of oxygen, O_{18} to trace the path of the oxygen, from water to gaseous molecular oxygen. This reaction is catalysed by a reactive centre in photosystem II containing four manganese ions.

The reaction begins with the excitation of a pair of chlorophyll molecules similar to those in the bacterial reaction centre. Due to the presence of chlorophyll *a*, as opposed to bacteriochlorophyll, photosystem II absorbs light at a shorter wavelength. The pair of chlorophyll molecules at the reaction centre are often referred to as P680. When the photon has been absorbed, the resulting high-energy electron is transferred to a nearby pheophytin molecule. This flow of electrons is similar to that of the bacterial reaction centre. Two electrons are required to fully reduce the loosely bound plastoquinone molecule to QH_2 as well as the uptake of two protons.

The difference between photosystem II and the bacterial reaction centre is the source of the electron that neutralises the pair of chlorophyll *a* molecules. In the bacterial reaction centre, the electron is obtained from a reduced compound haem group in a cytochrome subunit.

A difference between photosystem II and the bacterial reaction centre is the source of the electron which neutralises the pair of pigment molecules. Once photoinduced charge separation has taken place, the P680 molecule carries a positive charge. P680 is a very strong oxidant and extracts electrons from two water molecules which are bound at the manganese centre directly below the pair. This centre, contains four manganese ions, a calcium ion, a chloride ion, and a tyrosine residue. Manganese is used because it is capable of existing in four oxidation states; Mn^{2+}, Mn^{3+}, Mn^{4+} and Mn^{5+}. Manganese also forms strong bonds with oxygen-containing molecules such as water.

Every time the P680 absorbs a photon, it emits an electron, gaining a positive charge. This charge is neutralised by the extraction of an electron from the manganese centre which sits directly below it. The process of oxidising two molecules of water requires four electrons. The water molecules which are oxidised in the manganese centre are the source of the electrons which reduce the two molecules of Q to QH_2.

Photosystem I

After the electron has left photosystem II it is transferred to a cytochrome b6f complex and then to plastocyanin, a blue copper protein and electron carrier. The plastocyanin complex carries the electron that will neutralise the pair in the next reaction centre, photosystem I.

As with photosystem II and the bacterial reaction centre, a pair of chlorophyll *a* molecules initiates photoinduced charge separation. This pair is referred to as P700. 700 is a reference to the wavelength at which the chlorophyll molecules absorb light maximally. The P700 lies in the centre of the protein. Once photoinduced charge separation has been initiated, the electron travels down a pathway through a chlorophyll á molecule situated directly above the P700, through a quinone molecule situated directly above that, through three 4Fe-4S clusters and finally to an interchangeable ferredoxin complex. Ferredoxin is a soluble protein containing a 2Fe-2S cluster coordinated by four cysteine residues. The positive charge left on the P700 is neutralised by the transfer of an electron from plastocyanin. Thus the overall reaction catalysed by photosystem I is:

$$Pc(Cu^{+}) + Fd_{ox} \overset{light}{\Rightarrow} Pc(Cu^{2+}) + Fd_{red}$$

The cooperation between photosystems I and II creates an electron flow from H_2O to $NADP^+$. This pathway is called the 'Z-scheme' because the redox diagram from P680 to P700 resembles the letter z.

Light-harvesting Complex

A light-harvesting complex is one or more polypeptide chains containing photosynthetic pigments, which surrounds a photosynthetic reaction centre and focuses light inward, towards its core. Many such proteins have an alpha solenoid tertiary structure.

Light Harvesting Complexes in Plants

Chlorophyll b and carotenoids are important in light harvesting complexes present in plants. Chlorophyll b is almost identical to chlorophyll a except it has a formyl group in place of a methyl group. This small difference makes chlorophyll b absorb light with wavelengths between 400 and 500 nm more efficiently. Carotenoids are long linear organic molecules which alternate single and double bonds along their length. Such molecules are called polyenes. Two examples of carotenoids are lycopene and â-carotene. These molecules also absorb light most efficiently in the 400 – 500 nm range. Due to their absorption region, carotenoids appear red and yellow and provide most of the red and yellow colours present in fruits and flowers.

The carotenoid molecules also serve a safeguarding function. Carotenoid molecules suppress damaging photochemical reactions, particularly those including oxygen, which exposure to sunlight can cause. Plants that lack carotenoid molecules quickly die upon exposure to oxygen and light.

The chlorophyll b and carotenoids present in the light harvesting complexes are referred to as accessory pigments. These accessory pigments are held inside the light harvesting proteins in a highly uniform fashion. The light harvesting complexes are cylindrical in form and come in two sizes:

> LH-1 which are large and completely surround the reaction centre and LH-2 which are smaller and are arranged in a ring structure around the LH-1 complex. The relative size and positioning of the proteins is shown in the image above. LH-2 is the most abundant membrane protein in plant chloroplasts. Each protein contains 7 chlorophyll a molecules, 6 chlorophyll b molecules and 2 carotenoid molecules.

Photochemistry of Chlorophyll

It is the primary act, the capture of light energy for chemical purposes, which makes photosynthesis the most important and the most fascinating of biological processes. Even if every enzyme, with its attendant substrate, product, and co-enzyme, which enters into the biochemical cycles of the living plant were known, we might still be completely ignorant of the primary process, which distinguishes photosynthesis from other biological processes. There is no known straightforward way by which the nature of the primary act can be determined. In fact, there are very few photochemical reactions of complex molecules in solution, whose primary act has been unequivocally determined. In view of the difficulty of the problem, it appears worth while to investigate many different photochemical and spectroscopic properties of chlorophyll in solution, in synthetic and natural aggregates of chlorophyll molecules, and in the intact cell.

The study of dilute solutions of chlorophyll is certainly insufficient to establish the nature of the primary act of photosynthesis. However, knowledge gained from such study should enable the student of photosynthesis to reject many otherwise plausible interpretations of the primary act. This discussion is limited to recent developments of the photochemistry of chlorophyll solutions. Irreversible photochemical reactions of chlorophyll and reactions photosensitised by chlorophyll have been excluded. Although the irreversible reactions are of intrinsic interest to organic chemistry, they have no direct relation to normal photosynthesis. An adequate discussion of photosensitised reactions would require more space than is warranted by their importance. Furthermore, such reactions, occurring in dilute homogeneous solutions, cannot serve as reasonable models of photosynthesis.

Isomers and Tautomers

Freed and Saucier have shown that a tautomeric form of chlorophyll is stable at low temperatures, and they have studied, at intermediate temperatures, the tautomeric equilibria. This tautomerism is fatty even at 150°K. The absorption maxima of the low-temperature forms are displaced to the red by about 300 A. The temperature at which the low and high-temperature forms are present in comparable amounts is a sensitive function of the isomeric form of the chlorophyll. For example, the spectra of chlorophylls *b* and *b'* are almost identical at room temperature. Reducing the temperature to 75°K. changes both alike, so that the spectra of their low-temperature forms are also identical. However, the temperatures at which the tautomers coexist in equal amounts are 180°C and 230°K., respectively, for chlorophylls *b* and *b'*.

Chlorophylls *a* and *b* form addition compounds with water, alcohols, amines, and other "bases." Similar adducts are formed by bases with metal-complexed porphyrins and chlorines, but not with pheophytins or metal-free porphyrins. These facts support Krasnovskii's contention that the point of addition is the metal atom of the pigment. Unlike the other pigments, the fluorescent yield of chlorophyll is strongly affected by the addition of the base. Solutions of chlorophyll in pure dry hydrocarbons, or similar solvents, are practically non-fluorescent. The stability constants of compounds of chlorophylls, porphyrins, and chlorines with a given base do not differ by a factor of more than seven. As measured by these stability constants the oxygen bases are abnormally strong relative to nitrogen bases.

At 80°C, a solution of chlorophyll in isopropylamine is reddish brown, but when the solution is warmed to ordinary temperatures it becomes green and exhibits its normal spectrum. Except for a slow side reaction at the higher temperatures, the process is strictly reversible. The absorption spectrum of the low-temperature solution is identical with that of the Molisch phase test intermediate. This spectrum is similar to, but not the same as, the spectrum of chlorophyll in its triplet state, as formed by flash illumination. Since there is good reason for the belief that the phase test intermediate is an ion, formed by the loss of a proton.

Properties of Chlorophyll in its Singlet Excited States

Chlorophyll, excited by the absorption of blue light, emits only the usual red fluorescence. The molecule goes from its second to its first excited singlet state by a radiationless process of high probability (internal conversion). Since visible fluorescence having a ўield as high as 10^8 is readily detectable and no blue or green fluorescence has been observed, the quantum yield, of this fluorescence is less than 10^{-11} second. The actual mean life, $T = <PT°$, where $r°$ is the intrinsic mean life, must be less than $10^{-3}\times10^{-8} = 10^{-11}$ second.

It is, therefore, most improbable that the second excited singlet state plays any direct role in the photochemistry of chlorophyll. In other words, the photochemical action of visible light should be independent of wavelength whenever chlorophyll is the only light-absorbing entity present. This does not necessarily apply to the absorption of ultraviolet light. Higher excited states may possibly enter directly into photochemical reactions, but such processes would not be related to normal photosynthesis.

The intrinsic mean life, T°, of the first excited state of chlorophyll may be calculated (4) from its integrated extinction coefficient for "red" light. The actual mean life, r, may be obtained (4) from measurements of the degree of polarisation of fluorescent light in viscous solvents, by the use of Perrin's equation. Since the maximum quantum yields of fluorescence are known (8) for chlorophylls *a* and *b*, the intrinsic life may be obtained from the actual life, as T° = *T/<P°*. Weil's value for r° *(b)* is based upon measurements made with benzyl alcohol as the solvent, assuming that <*p*° = 0.10. Disregarding the values, which Rabinowitch calculated from the early data of Prins, the values of the intrinsic mean lives of chlorophylls a and *b* fall in the ranges 3 to 1.3 × 10^{-8} and 4 to 1.6 × 10^{-8} second, respectively. The maximum quantum yield, >°, for the fluorescence of chlorophyll *a* is 0.24, and it is independent of the solvent (8). The yield for chlorophyll *b* varies with the solvent, being 0.11 for ether and 0.06 for methanol. The corresponding values for the actual mean lives are r°(o) 5 × 10^{-9} and r°(&) 3 × 10^{-9} second.

It is well established (4,10,11) that energy of electronic excitation can be readily exchanged between chlorophyll molecules, separated by distances as great as 50 A. This process, which is called induced resonance, leads to self quenching and concentration depolarisation, when it occurs between like molecules, and to fluorescence quenching and sensitised fluorescence when it occurs between unlike molecules. For chlorophyll solutions, self quenching is a quadratic function of concentration (4,10), <p_{max}./<P = 1 + km^2. It becomes appreciable at *m* = 2 × 10^{-8} and reaches its half value (<$p_m$ax./<P == 2) at about 1.5 × 10^{-2} *m*. It is not a function of the viscosity of the solvent, being practically identical in benzyl alcohol (4), acetone (10), and ethyl ether (10). Concentration depolarisation follows a first-order law (4), *p*°/*p* = 1 + *k'm*. The relative

polarisation, $p/p°$, equals one half at a concentration of 6×10^{-8} *m.* By a slight extrapolation of these data one would predict that <*p* and *p* should equal 0.5 per cent and 1 per cent, respectively, in a 0.1 *m* randomly oriented solution of chlorophyll in a highly viscous solvent. Since the concentration of chlorophyll is approximately 0.1 m in the grana it is interesting to compare these calculated values to the corresponding quantities, which have been observed for chloroplasts and algae; 3 per cent (Latimer) and 2 per cent (Arnold), respectively. This near agreement suggests that the chlorophyll molecules are randomly oriented (at least in two dimensions), which is surprising in view of the evidence indicating that the pigment molecules in the grana exist in some ordered orientation.

In solutions which contain chlorophyll *a* and 6, the fluorescence of 6 is quenched by *a* (10) and the fluorescence of *a* is sensitised by 6 (10,11). The probability of the transfer of excitation from *b* to *a* leading to the fluorescence of *a* reaches a maximum of 0.5 at about 10^{-8} *m,* when the concentrations of *a* and *b* are equal (10,11). Concentration quenching becomes dominant at higher concentration and reduces the probability of sensitised fluorescence to a negligible value at 10^{-2} *m.* Sensitised fluorescence is a well-known phenomenon in the intact plant.

The fluorescence of chlorophyll solutions is strongly quenched by oxidising agents (such as quinones, aryl nitro compounds, nitroso compounds, and oxygen), but is not detectably quenched by a wide variety of reducing agents (including thiourea, urethane, hydro-quinone, ascorbic acid, etc.). Quenching by the oxidising agents is chiefly a diffusional process. The polarisation of fluorescence increases with the degree of quenching in accord with Perrin's equation. The rate constants corresponding to the bimolecular quenching reaction can be calculated from the ratio of the Stern-Volmer quenching constant, k_{Qj} to the actual mean life of the excited state, *T.* A few such values for chlorophyll *a* at 25°C. In these calculations a value of 5×10^{-9} second has been assumed for *T*. The agreement is fairly good, but there is probably some static quenching, which is more apparent in viscous solvents.

Properties of Chlorophyll in its Triplet State

Kinetics studies of photochemical reactions sensitised by chlorophyll demonstrate that it is usually not the fluorescent (i.e., first, excited singlet) state but rather a metastable state of chlorophyll which is directly responsible for the photochemical action. When the reactant is present at high concentration in a fluid medium or exists as a thermally stable complex with the chlorophyll, the fluorescent state of the chlorophyll molecule can play a direct part in the photochemical reaction. Such reactions are, of course, accompanied by marked quenching of the fluorescence.

By analogy with fluoresce in and with aryl hydrocarbons, it is commonly assumed that the metastable state is the lowest triplet state of the molecule. Most unsaturated compounds exhibit phosphorescence when they are dissolved in glassy media. Recently Becker and Kasha have confirmed the existence of a weak, near infrared phosphorescence of chlorophyll *b*. The present status of the subject may be summarised as follows. No temperature-dependent phosphorescence (with the same wavelength distribution as fluorescence) has been observed for either chlorophyll under any conditions. No phosphorescence of any type has been observed for chlorophyll a. Chlorophyll b shows a temperature-independent phosphorescence (i.e., long-lived fluorescence) at 8050 A, which has a half-time of about 3×10^{-2} second and a quantum yield less than 10^{-8}. The observed wavelength corresponds to an energy difference between the ground and the metastable states of 33.0 kcal.

Organic molecules are phosphorescent only when they are dissolved in rigid media or adsorbed on solids. Therefore, the measurement of phosphorescence is not well adapted to the investigation of the chemistry of the triplet states of such molecules. This is especially true for chlorophyll, whose phosphorescence is very weak. It has been reasonably well established that the disappearance of the triplet state occurs, at least in part, by a bimolecular, self-quenching process in fluid solvents. It is known that molecular oxygen quenches the triplet state efficiently; presumably, at every encounter. Allylthiourea has little if any effect upon the mean life of the triplet state.

Reversible Photochemical Reactions of Chlorophyll

Anaerobic solutions of chlorophyll in methanol, ether, etc. but not in hydrocarbons show rapid reversible changes in their absorption spectra when they are strongly illuminated. Under experimentally realisable conditions, the maximum change in the absorption of monochromatic light is less than 1 per cent. Since the percentage change is proportional to the square root of the intensity of the absorbed light, we may conclude that the labile product is a pair of radicals or ions which reform chlorophyll by recombining. The absorption spectrum of this intermediate has been determined only very crudely, but appears to resemble that of the triplet state.

It was observed by Linschitz that chlorophyll, in a rigid solvent (EPA at liquid nitrogen temperatures) containing a quinone, is photochemically transformed into a product which is stable at low temperatures but reverts immediately to normal chlorophyll when the glass is allowed to warm to its softening point. The absorption spectrum of this product is, likewise, generally similar to that of the triplet state. The earlier finding of Krasnovskii that chlorophyll can be reversibly, photochemically reduced by ascorbic acid, when dissolved in pyridine, has been confirmed by A. S. Holt and E. Rabinowitch and by Evstigneev and Gavrilova (17b). The latter authors also observed this reaction in toluene with phenylhydrazine as the reducing agent.

Absorption Spectra and Deductions

Freed and Sancier have described, the presence of various solvates and temperature-dependent isomers of chlorophylls and related compounds. From these data, it was determined that if solvation occurs at the magnesium atom of chlorophylls, it also occurs at or near the two corresponding hydrogen atoms of pheophytin.

Russian workers emphasised that the difference in activation of fluorescence between pheophytin and chlorophyll was due to the fact that hydration of magnesium was necessary. Thus, the fluorescence of pheophytin, which contains no magnesium, was insensitive to the presence or absence of water vapour. With consideration of the work of Freed, it now may be emphasised that the presence of the magnesium may well cause the perturbation of the electronic system of the chlorophyll molecule necessary for the facts concerning fluorescence to become observable. Consequently, the observed results concerning the effects of the presence or absence of water on the fluorescence may be only coincidental facts of interest accompanying the real cause of the difference in the two molecules, that is, the effect of the presence or absence of magnesium in the molecules.

Emission Spectra and Deductions

The phosphorescence of chlorophyll b, first reported by Calvin and Dorough, is a *bona fide* emission occurring at 8650 A. An analog of chlorophyll, pheophorbide a, has a strong fluorescence and no phosphorescence. However, in the divalent copper complex of pheophorbide a, the fluorescence is completely quenched and only a strong phosphorescence is observed. Moreover, the phosphorescence occurs at 8675. The complete quenching of fluorescence in the metallophosphorbide and the like positions of the phosphorescent emissions indicate that the phosphorescences observed in the pheophorbide complex and in the chlorophyll are the lowest triplet to singlet emissions.

The energy available in the lowest excited level corresponding to the wavelength of the phosphorescent emission is about 33 kcal. per mole. The energy available in the lowest singlet state is approximately 43 kcal. per mole. Although the energy available in the singlet state is higher, a consideration of the amount of absorbed energy re-emitted from the singlet state de-emphasises this factor. Approximately 90 per cent of the absorbed energy is unaccounted for in terms of re-emission; 10 per cent of the absorbed energy reappears as fluorescence.

Fluorescence Spectra of Protochlorophyll, Chlorophylls c *and* rf, *and their Pheophytins*

The fluorescence spectra of pheophytins *a* and *b* as well as of pheophytin *d* are very similar to those of the chlorophylls from which they are derived except for a 2 to 13 m/z shift to longer wavelengths.

Acid-treated chlorophyll *c* and protochlorophyll, however, show larger shifts of the main fluorescence band, and a greatly increased height and corresponding shift of the second fluorescence band.

Chlorophyll-sensitised Photochemical Reactions in Vitro

The photochemical reactions sensitised by chlorophyll *in vitro* shows that the only reactions, which proceed with high quantum yields are exothermic or slightly endothermic more end other mi c reactions have very small yields because these require that the bulk of the excitation energy of the chlorophyll be stored as chemical energy. In the process of sensitisation, the excitation energy must be greater than the sum of the heat of activation and the chemically stored energy. However, the difficulty vanishes if one assumes that, for the first photochemical step, the energy of two quanta rather than of one quantum is utilised. The transfer of a hydrogen atom from the chlorophyll to a weak oxidant is an example. When chlorophyll is in the lowest triplet state (corresponding to approximately 30 kcal.), even the transfer to oxidants as strong as molecular oxygen or quinone in one act becomes energetically impossible. There is a possibility that one hydrogen atom may be transferred in two chemical steps. However, the assumption which best fits the kinetic data for reactions *in vivo* is that two quanta are utilised to excite *one* chlorophyll molecule into an excited triplet state, whereby between 60 and 70 kcal. become available for *one* photochemical act.

That can be achieved by the following steps: (/) Excitation of a chlorophyll molecule by light absorption into the first excited singlet state. Internal transition into the lowest metastable, long-lived triplet state. Excitation of the chlorophyll in its metastable triplet J. FKANCK state to the next higher triplet state by sensitised fluorescence. The energy needed for this process comes from a neighbouring chlorophyll molecule in the first excited singlet state. Utilisation of the energy difference between the excited triplet and the singlet ground state for photochemistry.

When chlorophyll absorbs red light it is raised to its first excited (or fluorescent) state, usually with some extra vibrational energy. This excess vibrational energy is quickly lost, bringing the electronically excited molecule into thermal equilibrium with its surroundings. There is a large probability (between $^8/_4$ and J) that the excited molecule will not emit a photon of fluorescence light but will be transferred by a process of internal conversion into its lowest triplet level. In this state, the life of the molecule will be long compared to its life in the fluorescent state. When two similar molecules are close together but not in actual contact, there is a certain probability that, if one is electronically excited, it can lose that energy and the energy of excitation will appear in the second molecule.

This can also occur when the molecules are unlike, if certain requirements are satisfied. An empirical requirement is that there should be a strong overlap between the emission band of the first (primarily excited) molecule and the absorption band of the second molecule. The probability of such a transition caused by sensitised fluorescence is great when there is a large overlap between the emission band and the absorption band. For this reason, one might expect a greater probability of such radiationless transfer between unlike molecules than between like molecules.

Because the visible absorption spectra of the triplet state and of the Molisch phase test intermediate of chlorophyll are similar, Franck predicted that the triplet state, like the phase-test intermediate, would have an appreciable absorption in the near infrared close to the red maximum of the singlet absorption. This prediction was verified by Linschitz and more recently by Fujimori. There is, therefore, a strong overlap between the fluorescence of singlet-state chlorophyll and the absorption of its lowest triplet state. This condition strongly favours a radiationless transition of the excitation energy from the first-excited singlet to the lowest triplet level of chlorophyll.

Evidence on the Enzymatic Nature of the Process

1. The long-lived luminescence increases with increasing illuminating intensity up to a certain value and thereafter remains essentially constant; i.e., it displays a typical saturation as does an enzymatic reaction.
2. At unphysiological temperatures the luminescence is reversibly destroyed if the heat treatment is brief. Exposure to 45° to 50°C. for a few minutes irreversibly destroys the luminescence.
3. Rate of luminescence as a function of temperature shows a typical enzymatic temperature optimum at 37°C (the temperature optimum for photosynthesis) with a heat of activation for the process of ca. 10 to 15 kcal. That this activation energy is characteristic of the luminescent substrate was established by measuring in a flow system the temperature dependence of the emission process after illumination at different temperatures. The intensity of light emission was strongly dependent on the measuring temperature and essentially independent of the temperature at which the plants were illuminated (between 0° and 40°C). That the formation of luminescent substrate does

involve some activation energy was shown by illumination at liquid nitrogen temperature, after which no luminescence is observable upon thawing.

4. Various metabolic inhibitors, including UV, light, produce strong changes in luminescence.

Decay Characteristics

1. The luminescence does not decay according to any simple kinetic formulation. About one-half of the luminescence disappears in 0.1 second, after which it decays more slowly.
2. At least two definite components are distinguishable:
   a. A fast decaying component of ca. 0.01 second duration that does not appear to saturate at high light intensity (note resemblance to Emerson-Arnold time).
   b. A slowly decaying luminescence that reaches saturation at moderate incident light intensity.
3. The time course of the luminescence induction period is different for different portions of the decay. The long-term decay possesses a marked induction maximum whereas the fast decay component shows a much less marked maximum and actually increases with time of illumination.

Relation to Photosynthesis

1. In intact plants the intensity of luminescence is influenced by the presence or absence of CO_2 and by anaerobic conditions CO_2 depresses luminescence, as might be expected if it removes photochemically produced reductant. Prolonged exposure to anaerobic conditions inhibits the luminescence.

In a parallel manner, the luminescence of chloroplasts is inhibited by the addition of hydrogen acceptors such as the Hill oxidants, ferricyanide, or quinone.

The luminescence is also exhibited by chloroplasts and saturates parallel to the Hill reaction rate as a function of incident light intensity. The ability of chloroplasts to luminesce disappears during prolonged storage of chloroplasts at? 20°C., as does the Hill reaction. However, lyophilised powders retain activity apparently indefinitely. With the exception of a gradual increase in the level of luminescence under certain conditions during the first few minutes of illumination chloroplasts do not exhibit any induction effects. By contrast, the luminescence of intact plants evidences striking fluctuations in intensity during the first few minutes of illumination. These induction effects are very similar in their time course to a number of other transients in photosynthesis, e.g., fluorescence, ATP concentration, O_2 liberation, and CO_2 fixation.

These facts strongly suggest the following conclusions:

The luminescence of green plants is due to an enzymatically catalysed recombination of early photoproducts in photosynthesis, probably the primary reducing and oxidising agent or the very next reactants in the sequence of O_2 liberation and CO_2 reduction.

2. Conditions which would be expected to cause an accumulation of these photoproducts increase the luminescence and, conversely, conditions which should promote utilisation or decrease production of these products decrease the intensity of luminescence.
3. The induction curves are due to changes in concentration of a number of intermediates in the chain of reactions leading from the photochemical events to CO_2 fixation, including the formation and utilisation of ATP.
4. The decay curves suggest that two molecular species (probably on the reductant side) are capable of eliciting luminescence. The short-lived component may be regarded as the primary photoproduct and the longer

lived component as a substance derived from the first and representing a later hydrogen or OH carrier in the photosynthetic sequence. It is interesting that the short- and long-lived components behave kinetically similarly to the various constants derived from flashing light experiments and thus mixtures of different life-time intermediates may be responsible for the discrepant results reported by various authors under differing conditions of illumination, flash duration, etc.

Decay of the Delayed Light Emission in Chlorella

The process of photosynthesis seems to be partially reversible; green plants emit light for some time after they have been illuminated. Strehler (this volume) has given reasons for believing that this delayed light production is the reverse of photosynthesis. The present paper is a study of the decay of the delayed light in *Chlorella*. In general, no simple relation has been found between the intensity of the delayed light and the time in the dark.

All measurements of the intensity of the delayed light were made with an RCA #6217 Photomultiplier, at the temperature of dry ice, and with a vibrating reed electrometer.

Chemiluminescence of Algae

Algae were irradiated with repeated pulses of blue light. Between these flashes, the chemiluminescence was measured. The time sequence was (in milliseconds): irradiate, 1.25; dark, 0.75; measure, 1.25; dark, 0.75. The intensity of the luminescence was determined with a 1P22 multiplier phototube, fitted with a red filter. The photosignal was amplified and recorded. At the present time, a quantum counting technique is being used to metre the photosignal. A 3-ml. sample of algae was placed in a cell having a sintered glass bottom through which various gases could be passed another variance from the technique of Strehler and Arnold. *Chlorella pyrenoidosa* were used in the work reported here. Irradiation was also undertaken with green, yellow and red light (with and without admixtures of blue light). Within the uncertainties introduced due to light leak and other experimental difficulties, the results are qualitatively the same as those obtained with blue light. As Arnold and co-workers have shown, the spectral distribution of chemiluminescence is experimentally indistinguishable from that of the fluorescence of chlorophyll but is several orders of magnitude less intense.

The conclusion that the chemiluminescence is least intense when photosynthesis is proceeding with great facility is bolstered by studies made with various poisons, inhibitors, and narcotics.

Photochemical Part of Photosynthesis

The fluorescence yield of chlorophyll *a* in green plants is low and, very roughly speaking, constant over the range of illumination intensities from that which compensates respiration to the one which produces saturation. When secondary processes involving molecular oxygen are avoided, the fluorescence yield remains unaltered well into the region of saturation. This indicates that the processes principally responsible for the limitation of the fluorescence yield are those associated with the excitation of the chlorophyll molecules rather than with the utilisation of the excitation energy for photochemical purposes. The transition of the excited chlorophyll molecule into the metastable triplet state is the principal cause of the low fluorescence yield. Since the energy of this state is not sufficient to transfer one hydrogen atom from water to the usual photosynthetic oxidant at a sufficiently great rate.

Possible Role of Cytochromes Haematin Compounds in the Metabolism of Photosynthetic Tissues

Two salient facts, established only recently, provide the basis for this discussion. One is that all photosynthetic tissues contain haematein compounds. The second concerns the effect of the light energy utilised in photosynthesis. This energy sets up a steady state in which the major change relative to the dark steady state is a shift in the direction of greater oxidation of onę or more haematein compounds.

In the case of green plants and algae, it is not difficult to limit the data to be considered to those known to be unique to photosynthetic tissue. In these organisms, the sites of photometabolism are well-defined subcellular particles ("chloroplasts") about which a large body of knowledge is steadily accumulating. Much less well defined are the corresponding entities hi the bacteria. This uncertainty is not serious in the case of the obligate photoanaerobes, because these systems growth is possible only by photosynthesis under strictly anaerobic conditions. In the facultative bacteria, there is evidence that cell particles much like chloroplasts exist hi at least two species, *Rhodospirillum rubrum* and *Rhodopseudomonas spher aides*.

There is a great deal of significance in the fact that haematein compounds are found both in green plant chloroplasts and in large amounts hi all varieties of the photosynthetic bacteria. This significance lies in the fact that, whereas the green plants, algae, and the photosynthetic bacteria cover practically the whole range of living metabolic patterns, they share no property except that of using light energy for growth. Furthermore, until now only two classes of compounds have been found in all photosynthetic systems. These are the photoactive pigment complexes the chlorophylls, carotenoids, etc. and the haematein compounds.

The significance of the observation that absorption of light energy results in oxidation of component haematins lies in the apparent magnitude of these changes, in both amount and duration, which dwarf changes in other components expected to participate in energy transfer during the early phase of photosynthesis.

In elaboration of these remarks this report will be concerned, first, with the data relating to distribution and nature of haematein compounds in photosynthetic tissue, and, second, with the observations which have been made on the steady state of oxidation of cell components during photosynthesis.

Occurrence and Nature of Haematein Compounds in Green Plant Chloroplasts

Two haematein compounds appear to be unique to the photosynthetic apparatus in green plants. One of these, cytochrome b, was the first chloroplast haematein compound to be detected. Its isolation and properties have been described exhaustively by Hill and his collaborators.

Cytochrome b is widely distributed in the higher plants and algae. It appears to be an integral part of the chloroplast structure inasmuch as it cannot be separated from the chloroplast unless organic solvents are used to split off lipid, Cytochrome b is estimated to account for at least one-third of the haematein of the chloroplast and is present in the ratio of 1 mole for approximately 400 moles chlorophyll. It has been possible to obtain a preparation purified some 300-fold over the state in which cytochrome b is initially extracted from the leaf, using ammoniacal ethanol. The yield of purified material is about 13 per cent of the haematein originally present in the crude extract.

Cytochrome b is classified as a "modified" cytochrome of the *c* type because of its spectrochemical characteristics, the spectra of the haemochromogen derived from it by reaction with alkaline cyanide or pyridine, and its relatively high potential. However, it will not function as a substrate for the classical cytochrome *c* oxidase. In fact, no system spectrochemically analogous to cytochrome a appears to be present in the chloroplast, although it exists in the cytoplasm outside the chloroplast along with normal cytochrome *c*. Oxidase activity responding to respiratory inhibitors has been reported in chloroplaste but such activity is not proof for the presence of cytochrome a. Cytochrome a will couple with cytochrome *c* non-enzymatically so that it can be oxidised slowly when incubated in air with cytochrome *c* oxidase and catalytic amounts of cytochrome *c*.

The other chloroplast haematein compound is a "6" type cytochrome, called by Hill. This haematein compound appears to be associated with the chloroplasts of a variety of plants. Like cytochrome b, it is bound in the chloroplast structure firmly enough to withstand extraction by aqueous solvents.

Cytochrome b and cytochrome be cover a range of redox potential which is more than half of the total required to span the difference between the hydrogen and oxygen electrodes at physiological pH. Thus, these two compounds provide the chloroplast with an electron transfer system which, in principle, could carry out oxidations in either direction over a large fraction of the physiological range.

Haematein Compounds of Facultative Photosynthetic Bacteria

The non-sulphur purple bacteria are classified at present into two genera, *Rhodospirillum* and *Rhodopsendomonas*. Representative species contain large amounts of soluble haematein compounds which can be described as modified "c" cytochromes. Like cytochrome b, all these haem proteins show high positive redox potentials. Again like cytochrome b, they are not substrates for a cytochrome a type oxidase, and they can be coupled non-enzymatically with cytochrome c.

The haem protein from *R. rubrum* has been studied most intensively. Its properties, which are typical of many "c" cytochromes in photosynthetic organisms.

The facultative bacteria yield these proteins readily upon treatment with warm trichloroacetic acid as in the Keilin-Hartree procedure, and can be purified in the same manner as mammalian cytochrome c. However, they are noticeably more labile than their mammalian analogs and can be obtained in better yield with less denaturation by use of less drastic methods. In *R. rubrum* and *R. spheroides*, most of the cytochrome "c" appears to be loosely bound and extractable. There also is a fraction which is tightly bound to the particles and resists extraction with aqueous systems just as do the chloroplast hematins. There is some evidence that cytochromes of the "6" type are present in the chromatophores of the facultative bacteria. However, only haematein compound has been characterised.

A remarkable new type of haematein compound, as yet unclassified, has been observed which is distributed generally in the facultative bacteria. This compound displays the spectroscopic properties of a compound of the myoglobin type and shows, as well, an ability to form a CO complex in the reduced form. Nonetheless, it yields haemochromogen with spectra like those formed from a typical cytochrome c. The redox potential lies in the region normally attributed to haemoglobins or cytochrome *b*. It is clear that this haematein compound cannot be classified (at least, so far) in any of the common categories of the iron porphyrin proteins.

The enzymatic properties of this compound are also remarkable. It can act as a substrate for either the bacterial or mammalian cytochrome *c* reductase. Since it is rapidly autoxidisable and forms a CO compound in the reduced state it could function as a terminal respiratory oxidase. Another interesting property of this compound is its ability to reduce the bacterial cytochrome *c* non-enzymatically. In this respect it behaves in a manner reminiscent of mammalian cytochrome *c* with the bacterial cytochrome *c*. In other words, it provides a sort of oxidase for the cytochrome *c* of the bacteria as well as for the low potential electron donors present in the oxidative system. There is no satisfactory name for this compound at present. It has been called variously "pseudohaemoglobin," "yellow pigment, "and "CO-binding pigment." The difficulty of naming and classifying it can be resolved only by precise determination of its functions in bacterial metabolism.

Haematein Compounas of Strictly Anaerobic Photosynthetic Bacteria

The green sulphur bacteria and the purple sulphur bacteria

make up two divisions of the group of photosynthetic bacteria. They are marked off from the other division, the non-sulphur purple bacteria, because they can grow only as photoanaerobes. In the absence of light, they cannot use energy derived aerobically or anaerobically for growth. Thus, they provide a unique opportunity for the study of the function of haematein compounds in systems uncomplicated by pathways for utilisation of energy other than that available in photosynthesis.

In the green bacteria, only one genus *Chlorobium* is recognised. The few species known are differentiated from each other by characteristic substrate requirements for growth. Three haematein compounds are known. The first to be isolated in cell-free extracts is an iron porphyrin protein from *C. limicola*. Two others have been detected in *C. thiosidfaticum*. It is seen that these pigments while resembling a cytochrome of the *c* type show redox potentials more characteristic of cytochromes of the *V* type.

One haematein compound, isolated and purified from the purple sulphur bacterium, *Chromatium*, strain D, has been studied. Purification beyond that achieved by exhaustive electrophoresis of crude ammonium sulphate fractions has not been obtained.

At least two protein components are present. One of these binds a haem group similar to that found in mammalian cytochrome c. There is non-haem iron apparently present in some as yet unspecified form. While the compound is slowly autoxidisable and exhibits spectroscopic behaviour reminiscent of cytochrome *c*, it has a low redox potential. Thus, it too, falls into no recognised single class of iron porphyrin protein but is a hybrid compound like the cytochrome of the green bacteria and the yellow pigment of the facultative bacteria.

No enzymic function has been found for this compound, just as in the case of cytochrome c. Cytochrome *c* reductase fails to activate the *Chromatium* pigment, although extracts of *Chromatium* contain reductase activity when cytochrome *c* is present as a substrate.

It is apparent that the photochemical apparatus in photosynthetic systems, regardless of overall metabolic pattern, contains haematein compounds. Not only do these haematein compounds appear to be unique but they also do not occur outside the photochemical apparatus under conditions where there is no close coupling between normal respiration and photosynthesis.

The Function of Haematein Compounds in Photosynthesis

In green plant photosynthesis, the net result of the photochemical act is the transport of four electrons against a potential drop of 1.2 volts, which is the difference between the potentials of the hydrogen and oxygen electrodes at physiological pH. If the unitary theory of photosynthesis is adhered to, a similar process may be supposed to occur in bacterial photosynthesis, although no oxygen evolution concomitant with reduction at the level of the hydrogen electrode can be demonstrated.

In 1039, Hill proposed that the initial phase of the photochemical process did not push electrons all the way from the oxygen potential to the hydrogen potential, but only part way. A back oxidation was invoked to provide the additional energy storage. He suggested that the substrate for the back oxidation was a part of the reduced material formed in the initial photochemical act, and the H' acceptor was a part of the oxygen liberated by the partial movement of electrons away from the potential of oxygen. He also raised the possibility that the transport system for this movement of electrons consisted of respiratory pigments similar to, but not identical with, those normal to respiratory systems. It was inferred that this photorespiratory system was separated in the cell, spatially or otherwise, so that it could function in close coupling with the forward reaction of light absorption and act independently of normal respiration. At the time of these proposals, no basis existed for the notion of chloroplast respiratory pigments. As we have seen, however, there is now considerable experimental backing for the existence of such entities.

The notion of and back oxidation has intrigued many writers and has appeared in one form or another too often to recount here. Davenport and Hill have elaborated the concept of a back oxidation coupled to the photochemical process in terms of the haematein compound, cytochrome b, and some other oxidising agent which is reduced in the photochemical act. They point out that the potential for cytochrome b lies at a point more negative than that of the oxygen electrode by an amount which, for the movement of four electrons, is precisely equivalent to the energy of one quantum in the characteristic red emission spectrum of chlorophyll. The next drop in potential brings one to the region at the limit of the Hill chloroplast reaction and to the potential of cytochrome 6. Here, again, the potential difference results in an energy change for four electrons corresponding to the energy in one quantum of red light.

Gaffron and Rosenberg have summarised the experimental findings which indicate that no back oxidation occurs involving molecular oxygen as such. However, *pre-occupation with oxygen as the H acceptor in such a process is not requisite to the Hill postulate.* The oxidising system generated in the light can be a substance other than oxygen, such as a complex organic free radical which produces oxygen irreversibly and partly back reacts in the manner suggested by Hill.

It is fortunate that techniques now exist for making a start in determining the state of cell constituents in intact cells during metabolism. These are dynamic spectrophotometric methods, which have been employed by Duysens and others, and have reached a notable degree of development in the hands of Britton Chance and his co-workers.

Duysens first observed that, in intact cell suspensions of *R. rubrum,* there were changes in optical density brought about by a transition from a steady state in the dark to a steady state in the light. These changes coincided with the difference spectrum between oxidised and reduced forms of a cytochrome type compound or compounds. Vernon and Kamen had shown about the same time that the *R. rubrum* cytochrome *c* could be photochemically oxidised by air in an enzymic reaction, and that compounds at potentials more negative than that for the cytochrome component were not reactive. Shortly, thereafter, Chance and Smith conducted a more exhaustive spectrophotometric analysis of the action spectra exhibited by *R.rubrum* cell suspensions under a variety of experimental conditions.

In agreement with the other workers, they found that the net effect of illumination was an overall oxidation of one or more haematein compounds. In this facultative bacterium, which displays competition between dark aerobic oxygen uptake and light anaerobic metabolism, the effect of light under anaerobic conditions was qualitatively the same as that in the dark when oxygen was admitted.

Chance and Smith have proposed a scheme in which the haematein compounds *(R. rubrum* cytochrome *c* and yellow pigment) act as a bridge for electron transfer between the products of photolysis and the systems concerned with reduction of CO_2 and other substrates. Their basic assumption is that the haematein chain which reacts with oxygen can also react with the products of photolysis. They place the yellow pigment at the end of the respiratory chain. The redox potential in the isolated form appears to be negative by more than 100 mv. compared to the cytochrome *c* component. This fact would seem to disagree with the notion of the yellow pigment as a terminal oxidase. However, it has not been established that the yellow pigment has not undergone some modification during the procedures employed for extraction.

Kamen and Vernon have shown that, in *R. rubrum,* the reductase activity far exceeds the oxidase activity, using bacterial cytochrome *c* or mammalian cytochrome *c* as substrate. Hence the cytochrome *c* should be largely in the reduced state in the dark under aerobic conditions, as well as anaerobic conditions. Chance and Smith find that addition of phenyl mercuriacetate causes a rapid oxidation of the cytochrome c. This is in agreement with the known inhibition of reductase activity by this agent. In the absence of this inhibitor, they suppose that the photo-oxidant can react with the yellow pigment and thus pull electrons away from the cytochrome in a manner analogous to that observed by Vernon and Kamen for the coupled oxidation of cytochrome *c* by air in the presence of yellow pigment.

These latter workers have also shown that the photoenzymic oxidation of cytochrome *c* is not inhibited by cyanide as is the oxidase of the aerobic system in the dark. Similarly, Chance and Smith note that the light-induced oxidation is not affected by cyanide, and in addition is insensitive to CO. The isolated pigment has been described previously as forming a CO-complex, but apparently this compound may not be effective in preventing light oxidation because of the well-known dissociation of such haem-CO compounds by light. Both groups of workers postulate a competition between the photolytic oxidant and oxygen, with the site of the competition in the haematein chain. Chance and Smith propose that the shift to the oxidised state is the result of the preferential attack of the oxidase by the photo-oxidant.

Kamen and Vernon have observed that the photo-oxidation in air catalysed by their photo-oxidase proceeds at a rate at light saturation (which is attained at rather low light intensity) some five to ten times faster than the rate of the dark oxidation under the same conditions. They argue by analogy from this observation that competition occurs at the cytochrome *c* level. In this scheme, the bacterial cytochrome would act as the substrate for the two oxidative systems and the light oxidase would compete successfully for the cytochrome c during illumination. The shift from reduced to oxidised cytochrome, on their viewpoint, results from the fact that while the dark oxidase cannot keep up with the reductase in the absence of light, the increased rate of oxidation possible with the light oxidase favours a greater degree of oxidation in the light.

The complexities which arise in the interpretation of shifts in oxidation states in facultative tissues should be obviated by similar experiments in tissues of strictly photoanaerobic character. A start has been made by Olsen and Chance with *Chromatium*. Their preliminary findings again indicate that practically the entire action spectrum found can be accounted for in terms of photo-oxidation of the haematein system as isolated and characterised by Newton and Kamen, described above. This finding, even though preliminary, provides strong evidence that the haematein system of photosynthetic tissue is coupled to the photochemical act, because no other function for the haematein system is available in such a tissue. The bacterium cannot use a chemosynthetic oxidation apart from photometabolism to any useful purpose, and in fact, is inactivated and finally rendered non-viable by air.

A number of observations on chloroplasts and intact green plant tissues have been reported which parallel those recorded for the bacteria.

Change in Absorption at One Wavelength

The time course of the changes in absorption appeared to depend upon the suspension medium and the intensity of the exciting light.

As we know the changes in optical density of a suspension of *Rhodospirillum rubrum is* strain 4 at 430 HIM in anaerobic peptone. For intensities of the order of magnitude at which saturation of photosynthesis just occurs, the absorption decreases upon illumination graphs until a steady-state value is reached within a minute; upon darkening the change is reversed. At very high intensities more complicated curves were observed. A decrease in absorption *a* was followed by an increase b, which in its turn was followed by a decrease *c*. Upon darkening a decrease *d* occurred followed by an increase *e*.

If the light was left on, the decrease c went on for minutes. In the dark this decrease was slowly reversed. The much faster changes a, *b*, *d* and the fast part of *e* took place also after long illumination had produced an appreciable, not yet reversed, "bleaching" c.

It seems that the changes a and the fast part of *e* are caused by the same pigment as the changes a and *e,b* and *d* are presumably caused by a different pigment. Since the change *c* is slow, it is probably not caused by a photosynthetic catalyst, and, since the other changes seem to take place independently of *c*, it may be left out of consideration in the discussion of the changes *a, b, d* and the fast part of *e*.

Different Spectra

Under most experimental conditions the time course of the change in absorption is a monotonously increasing or decreasing function of the time, which approaches a steady-state value. If this steady-state value (at a certain constant exciting intensity) is plotted as a function of the wavelength of the measuring light, a "difference" spectrum is obtained.

The shapes of the difference spectra of *Rhodospirillum* appeared to depend on the suspension medium.

Three different spectra are represented:

1. The difference spectrum in anaerobic peptone.
2. That in aerobic distilled water.
3. That in anaerobic distilled water.

The difference spectrum in anaerobic peptone is similar to (albeit not identical with) the difference of the absorption spectra of oxidised and reduced cytochrome c. This indicates that upon illumination a cytochrome becomes more oxidised. Until its identity has been established, we shall call this cytochrome "cytochrome 428" after the wavelength of the maximum in its difference spectrum.

In aerobic distilled water there is a positive maximum (increase in absorption) at 432 m/i. This is not caused by reduction of cytochrome 428 or of *Rhodospirillum* cytochrome c, since the band is too broad and occurs at another wavelength. In the infrared, there is a decrease at 880 m/i and an increase at 790 m/. The infrared part of the spectrum is probably caused by a change in bacteriochlorophyll. Since this type of difference spectrum occurs in the presence of oxygen or oxidising conditions and disappears under reducing conditions, the change may be caused by an oxidation of bacteriochlorophyll. If a small part of the bacteriochlorophyll takes part in this reaction, then the difference spectrum indicates that the 880-m/ peak is shifted to 790 m/i and the near ultraviolet band to 430 m/i. These shifts may be interpreted as indicating the dehydrogenation of one of the two reduced pyrrole nuclei in bacteriochlorophyll.

In anaerobic distilled water the negative maximum occurs at 420 m/i, suggesting the oxidation of *Rhodospirillum rubrum* cytochrome c, a pigment isolated by Vernon and Kamen. The hump at about 430 m/x suggests the oxidation of cytochrome 428.

The change at 430 is saturated at a much lower intensity than the change at 420 m/i. Thus the peak at 420 and the hump at 430 m/i must be caused by different pigments a conclusion consistent with the suggestion that the changes at 430 and 420 are mainly caused by cytochromes 428 and c, respectively. Thus cytochrome 428 seems to attain saturation at a lower intensity than cytochrome c, it may further be concluded that in this experiment oxidation of cytochrome *c* is presumably not caused by oxidised cytochrome 428.

Conclusions

The experiments reported indicate that cytochrome 428 participates in photosynthesis, since its oxidation-reduction state is shifted considerably towards the oxidised side upon onset of photosynthesis. Also, cytochrome *c* can be oxidised by light, but this oxidation was observed only under slightly unphysiological conditions. It is possible, as Chance and Smith concluded, that it occurs to a small extent also in photosynthesising cells. This conclusion was based on the assumption that the minor band at 550 mi in the difference spectrum was caused by cytochrome *c*. It may, however, have been caused by a band of cytochrome 428. Vernon and Kamen extracted three cytochromes, one of which was c, from *Rhodospirillum* strain 1 and determined the absorption spectra in the oxidised and reduced state. The difference spectrum (oxidised minus reduced) of one of these cytochromes had a maximum at 428 m/i. However, the half-width of the band was 27 m/i, which is different from the 12-m/u half-width of cytochrome 428. The difference spectrum of the third cytochrome was also different from that of cytochrome 428.

The suggestion that cytochrome 428 was oxidised not only by light, but also by bubbling air through an anaerobic suspension, was confirmed by Chance and Smith.

The changes, interpreted above as caused by oxidation of bacteriochlorophyll, were observed only under conditions in which cytochrome 428 was in the oxidised state. It is possible, although not yet proved, that the changes *b* and *d* occurring at high exciting intensities in anaerobic peptone were also caused by oxidation of bacteriochlorophyll.

A small part of the bacteriochlorophyll is oxidised by light and simultaneously a reduced compound H, which may be a reduced pyridine nucleotide, is formed. Oxidised bacteriochlorophyll oxidises one or more reduced cytochromes. The oxidised cytochromes cause the oxidation of the substrate and of a small part of H, the last reaction leading to the formation of adenosine triphosphate (ATP), which assists in the reduction of CO_2. The reactions leading to the reduction of CO_2 may be analogous to those in algae. The postulated formation of ATP is in accordance with FrenkePs finding that illuminated extracts of strain 1 produce ATP from ADP and inorganic phosphate. It should be stressed that the scheme is provisional and incomplete.

Leaf Structure

In botany, a leaf is an above-ground plant organ specialised for photosynthesis. For this purpose, a leaf is typically flat (laminar) and thin, to expose the cells containing chloroplast (chlorenchyma tissue, a type of parenchyma) to light over a broad area, and to allow light to penetrate fully into the tissues. Leaves are also the sites in most plants where respiration, transpiration, and guttation take place. Leaves can store food and water, and are modified in some plants for other purposes. The comparable structures of ferns are correctly referred to as fronds. Furthermore, leaves are prominent in the human diet as leaf vegetables.

structurally complete leaf of an angiosperm consists of a petiole (leaf stem), a *lamina* (leaf blade), and stipules (small processes located to either side of the base of the petiole). The petiole attaches to the stem at a point that is called the "leaf axil". Not every species produces leaves with all of these structural components. In some species, paired stipules are not obvious or are absent altogether. A petiole may be absent, or the blade may not be laminar (flattened). The tremendous variety shown in leaf structure (anatomy) from species to species is presented in detail below under Leaf morphology. After a period of time (i.e. seasonally, during the autumn), deciduous trees shed their leaves. These leaves decompose into the soil.

A leaf is considered to be a plant organ, typically consisting of the following tissues:

1. An epidermis that covers thę upper and lower surfaces
2. An interior *chlorenchyma* called the mesophyll
3. An arrangement of veins (the vascular tissue).

External Features

The waxy cuticle at the top of the leaf is transparent, allowing light to enter for photosynthesis. It also stops transpiration for happening through the leaf, other than in the stomata.

The upper epidermal cells are also transparent to allow light to reach the chloroplasts for photosynthesis.

The lower epidermis has stomata, which allow gaseous

exchange to occur. Gaseous exchange being the intake of CO_2 and the release of O_2, which is essential for photosynthesis.

The leaf is thin, and has a very large surface area—making it ideal for diffusion and absorbsion. The larger the surface area, the more sunlight can be absorbed. The leaf is also thin so that the mesophyll cells are closer to the surface-reducing the diffusion distance of CO_2 from the surroundings to the mesophyll cells.

Internal Features

Vascular Tissue: This is to do with the xylem and the phloem. Remember that the xylem is always at the top of the leaf, and the phloem at the bottom. This is because the xylem transports water (needed for photosynthesis) to the leaf cells, and diffuses into the chloroplasts. The phloem transports food from the mesophyll cells to the rest of the plant. The xylem and phloem are vascular tissue that are situated in the leaf vein. The vein is supported by fibres (sclerenchyma), which keep the shape of the leaf, flat.

Mesophyll

There are two main types of Mesophyll cells, the palisade and spongy cells.

Palisade: It contains the most chloroplast, and are at the top of the leaf. Are closely packed, allowing more cells and therefore more chloroplasts to be near the surface. There are small intercellular spaces in between them which allows gaseous exchange to occur.

Spongy: They have less chloroplasts, and larger intercellular spaces. They also store carbohydrates made by photosynthesis. Dissolved carbohydrates diffuse into the phloem, to be transported to the rest of the plant.

Chloroplasts

These are the structures which contain chlorophyll, and this is where the photosynthesis actually happens. They have a double membrane, which is selectively permeable. This allows H_2O and CO_2 to diffuse into it, and O_2 to diffuse out of it.

Process of Photosynthesis

It is important to remember that O_2 comes from water, not CO_2.

Photosynthesis occurs in 2 phases—the light and dark reaction.

Light Reaction

Light "harvesting": In chloroplasts, there are thylakoids, which have Pigment molecules around them. These pigment molecules funnel light photons to the chlorophyll a molecules in the reaction centre. As the chlorophyll molecule absorbs this energy, its electrons move to a higher energy level – creating an 'excited' chlorophyll molecule. It can now pass this energy onto the chlorophyll molecule next to it, thus exciting it, and making itself normal again. Electrons are being transport through these 2 centres: Photosystem I (PS I)Red light (700nm) Photosystem II(PS II)Orange light (680nm)

Election Transport: This transfer of elections results in ATP and NADPH being formed. PS I is responsible for the creation of NADPH. When light is absorbed at PS I, an election is displaced and transferred to an election acceptor. This acceptor transfers the electron to NADP, reducing it to NADPH. But if the chlorophyll at PS I keeps loosing electrons, it will need to replace them. PS II does this. When light hits the chlorophyll at PS II, an electron is displaced and is sent, via electron carriers, to PS I. Thus giving PS I its lost electron back. But now the PS II has lost an electron, how does it get it back? The photolysis of water is the answer. Water is split and an electron donated to PS II. Oxygen is given off too – this being 1 of the products of photosynthesis. ATP is also formed by the transfer of electrons, and is made by ADP and an inorganic phosphate. The flow of electrons from PS II to PS I creates hydrogen ions (H^+), and a steep concentration gradient. The passage of H^+ out of the thylakoids, provides the energy for ATP to be synthesised in the presence of ATPase. This is called non-cyclic photophosphorylation. The H^+ ions also provide the hydrogen for reducing NADP to NADPH.

So the light reaction creates energy by phosphoralyation–producing ATP, and by adding electrons to NADP to create NADPH.

Dark Reaction

This reaction occurs in the stroma of the chloroplast. This stage is mainly concerned with the synthesising of carbohydrates from CO_2. This occurs in a number of steps, with different enzymes along the way.

The Calvin Cycle: Firstly CO_2 is combined with RuBP (a 5 carbon organic compound). The enzyme RuBP carboxylase is needed here. This "fixes" the CO_2. Thus a 6 carbon molecule is created, which is very unstable, so it splits immediately into 2 molecules (3 carbons each), of glycerate 3-phosphate(GP). GP is then reduced to GALP, a 3 carbon sugar. This reduction used hydrogen from NADPH. Energy for this comes from NADPH and ATP. The 3 carbon sugar is built up to a 6 carbon sugar, which can be converted to starch. Some of the 3C sugar isn't converted into 6C, but goes into the regeneration of RuBP, for the continuation of this cyclic process.

Factors Affecting the Rate of Photosynthesis

It can be affected by many things, including:

1. Sunlight—its Intensity and wavelength.
2. Temperature
3. CO_2 and O_2 —availability
4. Any factor that influences the production of chlorophyll, enzymes, or energy carriers (ATP and NADPH).

The rate of photosynthesis will always correspond to that factor which is in least supply. Its rate will only increase when its supply is increased.

Sunlight

Generally the more light there is, the more photosynthesis occurs. This is true up to a point, where the plant has reached maximum photosynthesis levels, and so any increase in light intensity will not affect the plant further.

Temperature

Generally higher temperatures are better than cold temperatures for photosynthesis. However if it is humid, and the air is saturated with water vapour, photosynthesis is limited. This is because the water vapour being expelled cannot leave and go into the air, as there isn't a concentration difference for it to diffuse across. So the water vapour sits in the leaf, stopping photosynthesis.

CO_2

Basically the more CO_2 in the air, the better the rate of photosynthesis. Of course CO_2 is taken in during the daytime, and if its content was less than normal, photosynthesis couldn't happen normally. Fertilizers increase the number of soil fungi and bacteria, thus increasing the levels of CO_2 from their respiration and decomposition processes. So fertilizers are good news for plants.

Water and CO_2

Plants need water for photosynthesis. If they lack it, they wilt. When they have a deficiency of water, their stomata close, so CO_2 cannot diffuse into the leaves. So CO_2 is also the limiting factor, although it is as a result of lack of water.

Veins

The veins are the vascular tissue of the leaf and are located in the spongy layer of the mesophyll. They are typical examples of pattern formation through ramification. The pattern of the veins is called venation.

The veins are made up of:

- Xylem, which brings water from the roots into the leaf.
- Phloem, which usually moves sap out, the latter containing the glucose produced by photosynthesis in the leaf.

The xylem typically lies over the phloem. Both are embedded in a dense parenchyma tissue, called "pith", with usually some structural collenchyma tissue present.

External leaf characteristics (such as shape, margin, hairs, etc.) are important for identifying plant species, and botanists have developed a rich terminology for describing leaf characteristics. These structures are a part of what makes leaves determinant, they grow and achieve a specific pattern and shape, then stop. Other plant parts like stems or roots are non-determinant, and will usually continue to grow as long as they have the resources to do so. Classification of leaves can occur through many different designative schema, and the type of leaf is usually characteristic of a species, although some species produce more than one type of leaf. The longest type of leaf is a leaf from palm trees, measuring at nine feet long.

Basic Leaf Types

- Ferns have fronds.
- Conifer leaves are typically needle-, awl-, or scale-shaped
- Angiosperm (flowering plant) leaves: the standard form includes stipules, a petiole, and a lamina.
- Lycophytes have microphyll leaves.
- Sheath leaves (type found in most grasses).
- Other specialised leaves (such as those of *Nepenthes*)

Arrangement on Stem

Different terms are usually used to describe leaf placement (phyllotaxis):

- *Alternate:* Leaf attachments are singular at nodes, and leaves alternate direction, to a greater or lesser degree, along the stem.
- *Opposite:* Leaf attachments are paired at each node; decussate if, as typical, each successive pair is rotated 90° progressing along the stem; or distichous if not rotated, but two-ranked (in the same geometric flat-plane).
- *Whorled:* Three or more leaves attach at each point or node on the stem. As with opposite leaves, successive whorls may or may not be decussate, rotated by half the angle between the leaves in the whorl (i.e., successive whorls of three rotated 60°, whorls of four rotated 45°, etc.). Opposite leaves may appear whorled near the tip of the stem.
- *Rosulate:* Leaves form a rosette

As a stem grows, leaves tend to appear arranged around the stem in a way that optimises yield of light. In essence, leaves form a helix pattern centred around the stem, either clockwise or counterclockwise, with (depending upon the species) the same angle of divergence. There is a regularity in these angles and they follow the numbers in a Fibonacci sequence: 1/2, 2/3, 3/5, 5/8, 8/13, 13/21, 21/34, 34/55, 55/89. This series tends to a limit of 360° x 34/89 = 137.52 or 137° 30', an angle known mathematically as the golden angle. In the series, the numerator indicates the number of complete turns or "gyres" until a leaf arrives at the initial position. The denominator indicates the number of leaves in the arrangement. This can be demonstrated by the following:

- alternate leaves have an angle of 180° (or 1/2),
- 120° (or 1/3): three leaves in one circle,
- 144° (or 2/5): five leaves in two gyres, and
- 135° (or 3/8): eight leaves in three gyres.

The fact that an arrangement of anything in nature can be described by a mathematical formula is not in itself mysterious.

Mathematics are the science of discovering numerical relationships and applying formulae to these relationships. The formulae themselves can provide clues to the underlying physiological processes that, in this case, determine where the next leaf bud will form in the elongating stem.

Divisions of the *Lamina*

Two basic forms of leaves can be described considering the way the blade is divided. A simple leaf has an undivided blade. However, the leaf shape may be formed of lobes, but the gaps between lobes do not reach to the main vein. A compound leaf has a fully subdivided blade, each leaflet of the blade separated along a main or secondary vein. Because each leaflet can appear to be a simple leaf, it is important to recognise where the petiole occurs to identify a compound leaf. Compound leaves are a characteristic of some families of higher plants, such as the Fabaceae. The middle vein of a compound leaf or a frond, when it is present, is called a rachis.

- *Palmately Compound:* Leaves have the leaflets radiating from the end of the petiole, like fingers off the palm of a hand, e.g. *Cannabis* (hemp) and *Aesculus* (buckeyes).
- *Pinnately Compound:* Leaves have the leaflets arranged along the main or mid-vein.
- *Odd Pinnate*: With a terminal leaflet, e.g. *Fraxinus* (ash).
- *Even Pinnate:* Lacking a terminal leaflet, e.g. *Swietenia* (mahogany).
- *Bipinnately Compound:* Leaves are twice divided, the leaflets are arranged along a secondary vein that is one of several branching off the rachis. Each leaflet is called a "pinnule". The pinnules on one secondary vein are called "pinna", e.g. *Albizia* (silk tree).
- *Trifoliate*: A pinnate leaf with just three leaflets, e.g. *Trifolium* (clover), *Laburnum* (laburnum).
- *Pinnatifid*: Pinnately dissected to the midrib, but with the leaflets not entirely separate, e.g. *Polypodium*, some *Sorbus* (whitebeams).

Characteristics of the *Petiole*

Petiolated leaves have a petiole. Sessile leaves do not: the blade attaches directly to the stem. In clasping or decurrent leaves, the blade partially or wholly surrounds the stem, often giving the impression that the shoot grows through the leaf. When this is actually the case, the leaves are called "perfoliate", such as in *Claytonia perfoliata*. In peltate leaves, the petiole attaches to the blade inside from the blade margin.

In some *Acacia* species, such as the Koa Tree (*Acacia koa*), the petioles are expanded or broadened and function like leaf blades; these are called phyllodes. There may or may not be normal pinnate leaves at the tip of the phyllode.

A stipule, present on the leaves of many dicotyledons, is an appendage on each side at the base of the petiole resembling a small leaf. Stipules may be lasting and not be shed (a stipulate leaf, such as in roses and beans), or be shed as the leaf expands, leaving a stipule scar on the twig (an exstipulate leaf).

- The situation, arrangement, and structure of the stipules is called the "stipulation".
- free,
- *adnate:* fused to the petiole base,
- *ochreate:* provided with ochrea, or sheath-formed stipules, e.g. rhubarb,
- encircling the petiole base,
- *interpetiolar:* between the petioles of two opposite leaves,
- *intrapetiolar:* between the petiole and the subtending stem.

Venation (arrangement of the veins)

There are two subtypes of venation, namely, *craspedodromous*, where the major veins stretch up to the margin of the leaf, and *camptodromous*, when major veins extend close to the margin, but bend before they intersect with the margin.

- *Feather-veined, reticulate:* The veins arise pinnately from a single mid-vein and subdivide into veinlets. These, in turn, form a complicated network. This type of venation is typical for dicotyledons.

- Pinnate-netted, penniribbed, penninerved, penniveined; the leaf has usually one main vein (called the mid-vein), with veinlets, smaller veins branching off laterally, usually somewhat parallel to each other; e.g. *Malus* (apples).
- Three main veins originate from the base of the lamina, as in *Ceanothus*.
- Palmate-netted, palmate-veined, fan-veined; several main veins diverge from near the leaf base where the petiole attaches, and radiate towards the edge of the leaf; e.g. most *Acer* (maples).
- *Parallel-veined, parallel-ribbed, parallel-nerved, penniparallel:* Veins run parallel most the length of the leaf, from the base to the apex. Commissural veins (small veins) connect the major parallel veins. Typical for most monocotyledons, such as grasses.
- *Dichotomous:* There are no dominant bundles, with the veins forking regularly by pairs; found in *Ginkgo* and some pteridophytes.

Leaf Morphology Changes within a Single Plant

- *Homoblasty:* Characteristic in which a plant has small changes in leaf size, shape, and growth habit between juvenile and adult stages.
- *Heteroblasty:* Characteristic in which a plant has marked changes in leaf size, shape, and growth habit between juvenile and adult stages.

Petiole

In botany, the petiole .(from Latin *peciolus* "little foot," diminutive of *pediculus* "foot stalk," itself a diminutive of *pes* "foot") is the small stalk attaching the leaf blade to the stem. The petiole usually has the same internal structure as the stem.

Outgrowths appearing on each side of the petiole are called stipules. Leaves lacking a petiole are called sessile, or clasping when they partly surround the leaf. Clasping leaves of the Poaceae have an extra structure called the ligule.

Phyllodes are modified petioles. In some plants, the petioles become flattened and widened, while the leaf itself becomes reduced or vanishes altogether. Thus the phyllode comes to serve the purpose of the leaf. Phyllodes are common in the genus *Acacia*, especially the Australian species. Sometimes, especially on younger plants, partially formed phyllodes bearing reduced leaves can be seen.

Stipules and Stipels

In many orders of angiosperms, mainly among dicotyledons, the leaf may possess one or, more commonly, two stipules. A stipule is an outgrowth of the lower zone (Unterblatt) of a young leaf, part of the leaf base.

Stipules are relatively easy to observe and diagnose when these appendages are attached to the base of the petiole (i.e., the proximal end), but often the stipules deceptively appear to arise from the stem (interpetiolar stipules; Examples: *Tithonia diversifolia* and a geranium), although technically they are not part of the stem. Always the best place to look for stipules is among the immature, developing leaves at the shoot tip, because in many plant species the stipules die, shrivel, and are abscised relatively soon after the leaf has finished it growth (Example: a lupine).

The ochrea is described as fused stipules that form a sheath around the stem. This is especially easy to observe in the broad-leaved species of the genus *Rumex*, which often has a persistent, membranous ochrea, and Family Magnoliaceae, in which they are deciduous (Examples in Magnoliaceae: *Magnolia grandiflora, M. sprengeri, Talauma*, and a tulip tree).

Stipule-like outgrowths may also be found at the base of leaflets on a compound leaf, in addition to stipules at the base of the petiole (rachis). Leaflet stipules are termed stipels or, less commonly, secondary stipules.

When stipules abscise, a stipular scar develops. In certain lineages, e.g., Family Magnoliaceae (Example: *Talauma hodgsonii*) or the genus *Ficus* (Family Moraceae), the stipule encircles the stem and thereby forms an annular scar, diagnostic on young woody stems.

Which Plants Have Stipules?

Although nobody has counted, probably less than a quarter of all known angiosperm species have stipules. Among the dicotyledons, stipules have been observed in approximately 140 families, although in nearly half of these only relatively few species possess stipules. There are many groups of angiosperms in which stipules have never been reported, so that when stipules are observed, they often help an investigator to guess its home family.

In certain families or orders, such as legumes (Examples: *Melilotus*, *Bauhinia*, and *Calliandra*) and roses (Examples: mountain mahogany and *Photinia*), stipules are characteristic and nearly always present in some form. Most species of the exceeding large family Rubiaceae have interpetiolar stipules (A typical example: *Galium*). The large stipule of sycamore species (*Platanus*) appears to surround the stem and is very leaflike. Families of the wideranging orders Malvales (including mallow, hibiscus, and Chinese bellflower) and related Urticales (ramie) and Euphorbiales have stipules. Stipules are uncommon among the monocotyledons, where they tend to be small and vestigial.

Forms and Functions

The original function of stipules is obscure, but may have been involved as protection for the emerging leaves (stipules enclosing young bud and young leaf emerging from stipule pair in *Exbucklandia*). Conspicuous stipules covering buds also can be observed in *Cunonia capensis*. In some plant groups, e.g., figs (*Ficus*), relatives of teak (Family Dipterocarpaceae), and certain mangroves (Family Rhizophoraceae), the stipule is well developed and appears to serve in bud protection. However, the majority of angiosperms living today show no disadvantage by lacking them, and in many the stipule is small and vestigial, without any obvious function.

There are a number of interesting functions demonstrated for stipules of certain plant species.

1. Stipules that are green may be leaflike (Examples: flowering quince, a geranium, and a sycamore), linear, threadlike, or reduced to minute scales, thereby requiring a hand lens to see. Plant biologists have assumed that green stipules are photosynthetic, but ecophysiologists generally have not determined how significant green stipules may be as temporary or persistent photosynthetic structures. Anatomical reports on photosynthetic stipules are exceedingly rare, so that we know little about the presence of stomates and the nature of the mesophyll in stipules.
2. If hardened with lignin, a stipule may become modified as a spine, i.e., a stipular spine. In many species, one stipular spine or a pair of stipular spines may be observed at each node, e.g., among legumes, and for legumes, the stipular spine is the typical type of shoot armature (Examples: coral tree, *Chloroleucon*, and a mimosa). Obviously, stipular spines were effective in their specific native habitat for many of these cases as an adaptation to reduce predation by vertebrate herbivores. Very large, hollow stipular spines in certain species of bull acacia (*Acacia*, Family Mimosaceae) serve as homes for ants, which defend the plant from all other herbivores. Stipular spines are extremely important forms of shoot armature for the succulent euphorbs (Family Euphorbiaceae; examples: crown-of-thorns, young stipular spines, and old stipular spines), most species of which occur in thorny scrub and woodland as well as desert habitats in Africa and

southern Asia. Stipular spines also are present on certain climbers.

3. In many groups of plants, stipules are short-lived, non-photosynthetic structures that abscise before the leaf is fully expanded (a legume). Such stipules tend to be non-green, often hyaline (an elm), off-white (ramie), yellow, or brownish, but there are some remarkable cases where the stipule is brilliantly coloured, especially with anthocyanins, red to purple vacuolar pigments (*Talauma hodgsonii*). A non-green stipule probably can be regarded as vestigial, lacking any function for the mature leaf and shoot but perhaps protecting the young leaf while it is at the growing tip.
4. Sometimes stipules and stipels have evolved as extrafloral nectaries, utilised to nourish mutualistic ants, or other types of glands.
5. In *Smilax*, a lianaceous monocotyledon, the stipule may develop as a tendril.
6. For whatever function, corky stipules are known from certain dryland shrubs, for example, the species of creosote bush (*Larrea* young stipules of *L. tridentata* and older stipules in *L. nitida*) in New World deserts and the tough-leaved evergreen species of *Ceanothus* in California chaparral.

Photosynthetic Energy Transformation

The energy that drives photosynthesis originates in the centre of the sun, where mass is converted to heat by the fusion of hydrogen. Over time, the heat energy reaches the sun's surface, where some of it is converted to light by black body radiation that reaches the earth. A small fraction of the visible light incident on the earth is absorbed by plants. Through a series of energy transducing reactions, photosynthetic organisms are able to transform light energy into chemical free energy in a stable form that can last for hundreds of millions of years (e.g., fossil fuels).

The photosynthetic process in plants and algae occurs in small organelles known as chloroplasts that are located inside cells. The more primitive photosynthetic organisms, for example oxygenic cyanobacteria, prochlorophytes and anoxygenic photosynthetic bacteria, lack organelles. The photosynthetic reactions are traditionally divided into two stages - the "light reactions," which consist of electron and proton transfer reactions and the "dark reactions," which consist of the biosynthesis of carbohydrates from CO_2. The light reactions occur in a complex membrane system (the photosynthetic membrane) that is made up of protein complexes, electron carriers, and lipid molecules.

The photosynthetic membrane is surrounded by water and can be thought of as a two-dimensional surface that defines a closed space, with an inner and outer water phase. A molecule or ion must pass through the photosynthetic membrane to go from the inner space to the outer space. The protein complexes embedded in the photosynthetic membrane have a unique orientation with respect to the inner and outer phase. The asymmetrical arrangement of the protein complexes allows some of the energy released during electron transport to create an electrochemical gradient of protons across the photosynthetic membrane.

Photosynthetic electron transport consists of a series of individual electron transfer steps from one electron carrier to another. The electron carriers are metal ion complexes and aromatic groups. The metal ion complexes and most of the aromatic groups are bound within proteins. Most of the proteins involved in photosynthetic electron transport are composed of numerous polypeptide chains that lace through the membrane, providing a scaffolding for metal ions and aromatic groups. An electron enters a protein complex at a specific site, is transferred within the protein from one carrier to another, and exits the protein at a different site.

The protein controls the pathway of electrons between the carriers by determining the location and environment of the metal ion complexes and aromatic groups. By setting the distance between electron carriers and controlling the electronic environment surrounding a metal ion complex or aromatic group, the protein controls pairwise electron transfer reactions. Between proteins, electron transfer is controlled by distance and free energy, as for intraprotein transfer, and by the probability that the two proteins are in close contact. Protein association is controlled by a number of factors, including the structure of the two proteins, their surface electrical and chemical properties and the probability that they collide with one another. Not all electron carriers are bound to proteins. The reduced forms of plastoquinone or ubiquinone and nicotinamide adenine dinucleotide phosphate (NADPH) or NADH act as mobile electron carriers operating between protein complexes. For electron transfer to occur, these small molecules must bind to special pockets in the proteins known as binding sites. The binding sites are highly specific and are a critical factor in controlling the rate and pathway of electron transfer.

The light reactions convert energy into several forms. The first step is the conversion of a photon to an excited electronic state of an antenna pigment molecule located in the antenna system. The antenna system consists of hundreds of pigment molecules (mainly chlorophyll or bacteriochlorophyll and carotenoids) that are anchored to proteins within the photosynthetic membrane and serve a specialised protein complex known as a reaction centre.

The electronic excited state is transferred over the antenna molecules as an exciton. Some excitons are converted back into photons and emitted as fluorescence, some are converted to heat, and some are trapped by a reaction centre protein. Excitons trapped by a reaction centre provide the energy for the primary photochemical reaction of photosynthesis — the transfer of an electron from a donor molecule to an acceptor molecule. Both the donor and acceptor molecules are attached to the reaction centre protein complex. Once primary charge separation occurs, the subsequent electron transfer reactions are energetically downhill.

In oxygenic photosynthetic organisms, two different reaction centres, known as photosystem II and photosystem I, work concurrently but in series. In the light photosystem II feeds electrons to photosystem I. The electrons are transferred from photosystem II to the photosystem I by intermediate carriers. The net reaction is the transfer of electrons from a water molecule to NADP+, producing the reduced form,

NADPH. In the photosynthetic process, much of the energy initially provided by light energy is stored as redox free energy (a form of chemical free energy) in NADPH, to be used later in the reduction of carbon.

In addition, the electron transfer reactions concentrate protons inside the membrane vesicle and create an electric field across the photosynthetic membrane. In this process the electron transfer reactions convert redox free energy into an electrochemical potential of protons. The energy stored in the proton electrochemical potential is used by a membrane bound protein complex (ATP-Synthase) to covalently attach a phosphate group to adenosine diphosphate (ADP), forming adenosine triphosphate (ATP). Protons pass through the ATP-Synthase protein complex that transforms electrochemical free energy into a type of chemical free energy known as phosphate group-transfer potential (or a high-energy phosphate bond) (Klotz, 1967).

The energy stored in ATP can be transferred to another molecule by transferring the phosphate group. The net effect of the light reactions is to convert radiant energy into redox free energy in the form of NADPH and phosphate group-transfer energy in the form of ATP. In the light reactions, the transfer of a single electron from water to NADP+ involves about 30 metal ions and 7 aromatic groups. The metal ions include 19 Fe, 5 Mg, 4 Mn, and 1 Cu. The aromatics include quinones, pheophytin, NADPH, tyrosine and a flavoprotein. The NADPH and ATP formed by the light reactions provide the energy for the dark reactions of photosynthesis, known as the Calvin cycle or the photosynthetic carbon reduction cycle. The reduction of atmospheric CO_2 to carbohydrate occurs in the aqueous phase of the chloroplast and involves a series of enzymatic reactions.

The first step is catalysed by the protein Rubisco (D-ribulose 1,5-bisphosphate carboxylase/oxygenase), which attaches CO_2 to a five-carbon compound. The reaction produces two molecules of a three-carbon compound. Subsequent biochemical reactions involve several enzymes that reduce carbon by hydrogen transfer and rearrange the carbon compounds to synthesise carbohydrates. The carbon reduction cycle involves the transfer and rearrangement of chemical bond energy.

In anoxygenic photosynthetic organisms water is not used as the electron donor. Electron flow is cyclic and is driven by a single photosystem, producing a proton electrochemical gradient that is used to provide energy for the reduction of NAD+ by an external H-atom or e-donor (e.g., H_2S or an organic acid) in a process known as "reverse electron flow". Fixation of CO_2 occurs via different pathways in different organisms.

Oxygenic Photosynthesis

In plants the photosynthetic process occurs inside chloroplasts, which are organelles found in certain cells. Chloroplasts provide the energy and reduced carbon needed for plant growth and development, while the plant provides the chloroplast with CO_2, water, nitrogen, organic molecules and minerals necessary for the chloroplast biogenesis. Most chloroplasts are located in specialised leaf cells, which often contain 50 or more chloroplasts per cell. Each chloroplast is defined by an inner and an outer envelope membrane and is shaped like a meniscus convex lens that is 5-10 microns in diameter, although many different shapes and sizes can be found in plants.

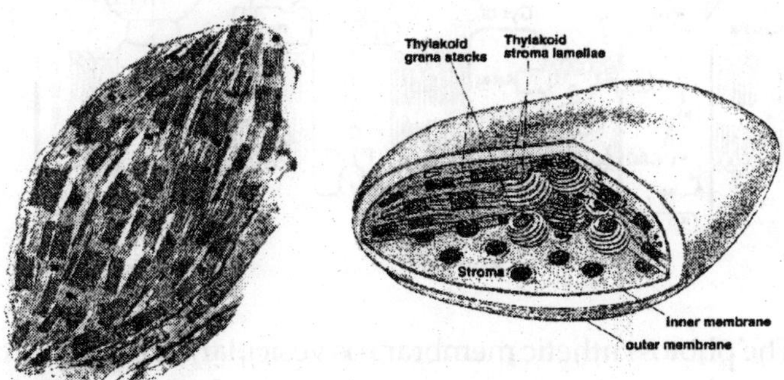

The inner envelope membrane acts as a barrier, controlling the flux of organic and charged molecules in and out of the chloroplast. Water passes freely through the envelope membranes, as do other small neutral molecules like CO_2 and O_2. There is evidence that chloroplasts were once free living bacteria that invaded a non-photosynthetic cell long ago. They have retained some of the DNA necessary for their assembly, but much of the DNA necessary for their biosynthesis is located in the cell nucleus. This enables a cell to control the biosynthesis of chloroplasts within its domain.

Inside the chloroplast is a complicated membrane system, known as the photosynthetic membrane (or thylakoid membrane), that contains most of the proteins required for the light reactions. The proteins required for the fixation and reduction of CO_2 are located outside the photosynthetic membrane in the surrounding aqueous phase. The photosynthetic membrane is composed mainly of glycerol lipids and protein. The glycerol lipids are a family of molecules characterised by a polar head group that is hydrophilic and two fatty acid side chains that are hydrophobic. In membranes, the lipid molecules arrange themselves in a bilayer, with the polar head towards the water phase and the fatty acid chains aligned inside the membrane forming a hydrophobic core.

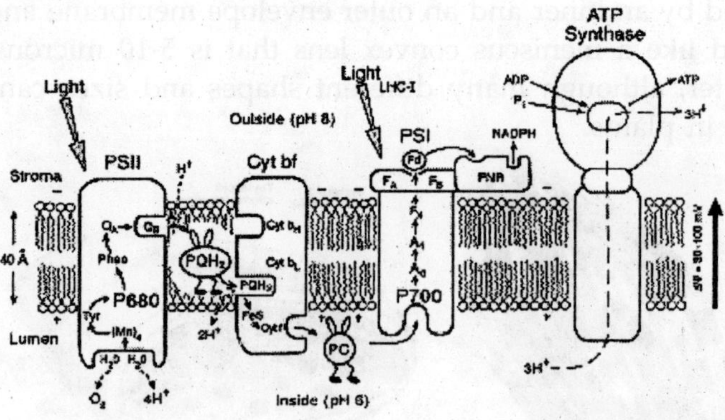

The photosynthetic membrane is vesicular, defining a closed space with an outer water space (stromal phase) and an inner water space (lumen). The organisation of the photosynthetic membrane can be described as groups of stacked membranes (like stacks of pita or chapati bread with the inner pocket representing the inner aqueous space), interconnected by non-stacked membranes that protrude from the edges of the stacks.

Experiments indicate that the inner aqueous space of the photosynthetic membrane is likely continuous inside of the chloroplast. It is not known why the photosynthetic membrane forms such a convoluted structure. To understand the energetics of photosynthesis the complicated structure can be ignored and the photosynthetic membrane can be viewed as a simple vesicle.

Light Absorption - the Antenna System

Plant photosynthesis is driven primarily by visible light (wavelengths from 400 to 700 nm) that is absorbed by pigment molecules (mainly chlorophyll *a* and *b* and carotenoids). The chemical structure of chlorophyll *a* molecule is shown in Fig.

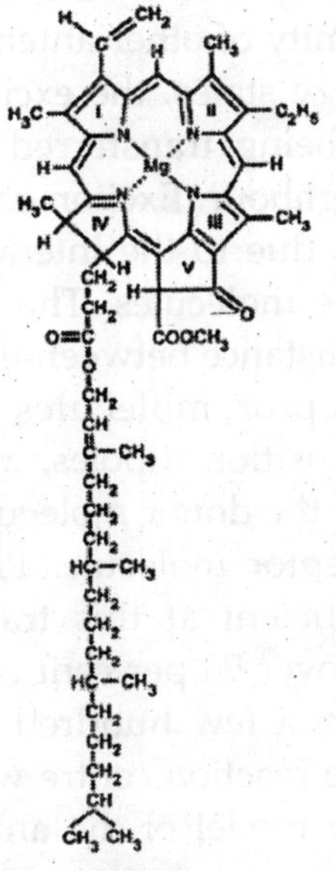

Chlorophyll *a*

In chlorophyll *b*, CH_3 in ring II is replaced by CHO group. Plants appear green because of chlorophyll, which is so plentiful that regions of the earth appear green from space. The absorption spectrum of chloroplast chlorophyll *a* and *b* and carotenoids along with the action spectrum of photosynthesis of a chloroplast. Light is collected by 200-300 pigment molecules that are bound to light- harvesting protein complexes located in the photosynthetic membrane.

The light-harvesting complexes surround the reaction centres that serve as an antenna. The three-dimensional structure of the light-harvesting complex shows that the protein determines the position and orientation of the antenna pigments. Photosynthesis is initiated by the absorption of a photon by an antenna molecule, which occurs in about a femtosecond (10-15 s) and causes a transition from the electronic ground state to an excited s ate.

Within 10-13 s the excited state decays by vibrational relaxation to the first excited singlet state. The fate of the excited state energy is guided by the structure of the protein. Because of the proximity of other antenna molecules with the same or similar energy states, the excited state energy has a high probability of being transferred by resonance energy transfer to a near neighbour. Exciton energy transfer between antenna molecules is due to the interaction of the transition dipole moment of the molecules. The probability of transfer is dependent on the distance between the transition dipoles of the donor and acceptor molecules (1/R6), the relative orientation of the transition dipoles, and the overlap of the emission spectrum of the donor molecule with the absorption spectrum of the acceptor molecule. Photosynthetic antenna systems are very efficient at this transfer process. Under optimum conditions over 90 per cent of the absorbed quanta are transferred within a few hundred picoseconds from the antenna system to the reaction centre which acts as a trap for the exciton. A simple model of the antenna and its reaction centre is shown in Fig.

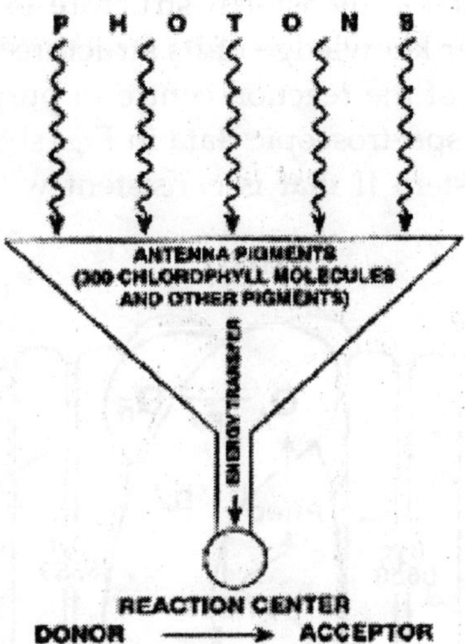

Primary Photochemistry — Photosystem II and Photosystem I Reaction Centres

Photosystem II uses light energy to drive two chemical reactions - the oxidation of water and the reduction of plastoquinone. The photosystem II complex is composed of more than fifteen polypeptides and at least nine different redox components (chlorophyll, pheophytin, plastoquinone, tyrosine, Mn, Fe, cytochrome b559, carotenoid and histidine) have been shown to undergo light-induced electron transfer (Debus, 1992). However, only five of these redox components are known to be involved in transferring electrons from H_2O to the plastoquinone pool — the water oxidising manganese cluster (Mn)4, the amino acid tyrosine, the reaction centre chlorophyll (P680), pheophytin, and the plastoquinone molecules, QA and QB. Of these essential redox components, tyrosine, P680, pheophytin, QA and QB have been shown to be bound to two key polypeptides that form the heterodimeric reaction centre core of photosystem II (D1 and D2). Recent work indicates that the D1 and D2 polypeptides also provide ligands for the (Mn)

4 cluster. The three-dimensional structure of photosystem II is not known. Our knowledge of its structure is guided by the known structure of the reaction centre in purple bacteria and biochemical and spectroscopic data in Fig. shows a schematic view of photosystem II that is consistent with current data.

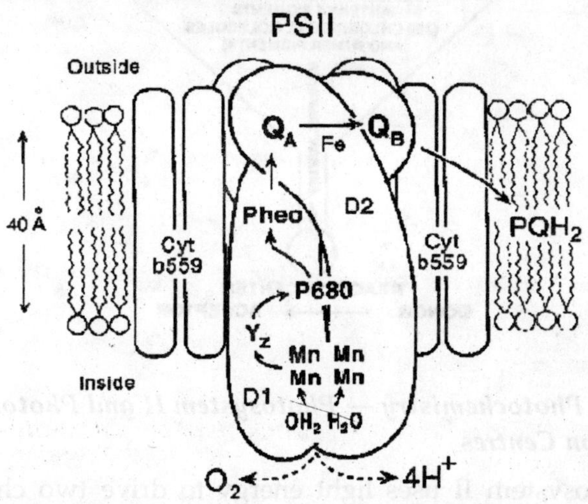

Photochemistry in photosystem II is initiated by charge separation between P680 and pheophytin, creating P680+/ Pheo. Primary charge separation takes about a few picoseconds. Subsequent electron transfer steps have been designed through evolution to prevent the primary charge separation from recombining. This is accomplished by transferring the electron within 200 picoseconds from pheophytin to a plastoquinone molecule (QA) that is permanently bound to photosystem II. Although plastoquinone normally acts as a two-electron acceptor, it works as a one-electron acceptor at the QA-site. The electron on QA- is then transferred to another plastoquinone molecule that is loosely bound at the QB-site. Plastoquinone at the QB-site differs from QA in that it works as a two-electron acceptor, becoming fully reduced and protonated after two photochemical turnovers of the reaction centre. The full reduction of plastoquinone requires the addition of two electrons and two protons, i.e., the addition of two hydrogen atoms. The reduced plastoquinone then debinds from the reaction centre and diffuses into the hydrophobic core of the membrane.

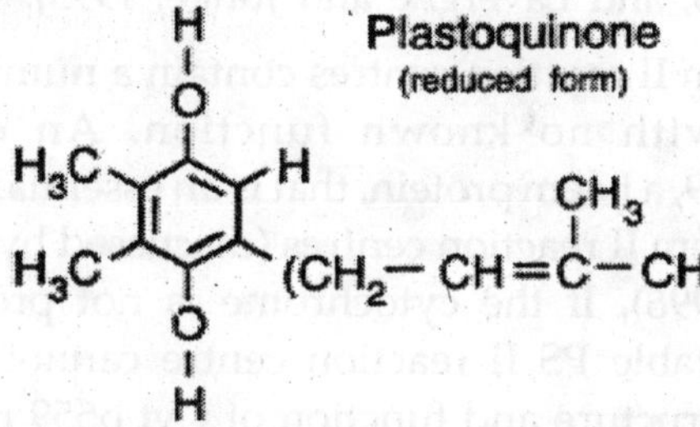

After which, an oxidised plastoquinone molecule finds its way to the QB-binding site and the process is repeated. Because the QB-site is near the outer aqueous phase, the protons added to plastoquinone during its reduction are taken from the outside of the membrane.

Photosystem II is the only known protein complex that can oxidise water, resulting in the release of O_2 into the atmosphere. Despite years of research, little is known about the molecular events that lead to water oxidation. Energetically, water is a poor electron donor. The oxidation- reduction mid-point potential (Em,7) of water is +0.82 V (pH 7). In photosystem II this reaction is driven by the oxidised reaction centre, P680+ (the mid-point potential of P680/P680+ is estimated to be +1.2 V at pH 7). How electrons are transferred from water to P680+ remains a mystery (Govindjee and Coleman, 1990). It is known that P680+ oxidises a tyrosine on the D1 protein and that Mn plays a key role in water oxidation. Four Mn ions are present in the water oxidising complex. X-ray absorption spectroscopy shows that Mn undergoes light-induced oxidation. Water oxidation requires two molecules of water and involves four sequential turnovers of the reaction centre. This was shown by an experiment demonstrating that oxygen release by photosystem II occurs with a four flash dependence. Each photochemical reaction creates an oxidant that removes one electron. The net reaction results in the release of one O_2 molecule, the deposition of four protons into the inner water phase, and the transfer of four electrons to the QB-site (producing two reduced plastoquinone molecules) (reviewed by Renger, 1993; Klein *et al.*, 1993; and Lavergne and Junge, 1993).

Photosystem II reaction centres contain a number of redox components with no known function. An example is cytochrome b559, a haem protein, that is an essential component of all photosystem II reaction centres (discussed by Whitmarsh and Pakrasi, 1996). If the cytochrome is not present in the membrane, a stable PS II reaction centre cannot be formed. Although the structure and function of Cyt b559 remain to be discovered, it is known that the cytochrome is not involved in the primary enzymatic activity of PS II, which is the transfer of electrons from water to plastoquinone. Why PS II reaction centres contain redox components that are not involved in the primary enzymatic reactions is a puzzling question. The answer may be found in the unusual chemical reactions occurring in PS II and the fact that the reaction centre operates at a very high power level. Photosystem II is an energy transforming enzyme that must switch between various high energy states that involve the creation of the powerful oxidants required for removing electrons from water and the complex chemistry of plastoquinone reduction which is strongly influenced by protons. In saturating light a single reaction centre can have an energy throughput of 600 eV/s (equivalent to 60,000 kW per mole of PS II). Operating at such a high power level results in damage to the reaction centre. It may be that some of the "extra" redox components in photosystem II may serve to protect the reaction centre. Photosystem II has another perplexing feature. Many plants and algae have been shown to have a significant number of photosystem II reaction centres that do not contribute to photosynthetic electron transport (e.g., Chylla and Whitmarsh, 1989). Why plants devote resources for the synthesis of reaction centres that apparently do not contribute to energy conversion is unknown.

The photosystem I complex catalyses the oxidation of plastocyanin, a small soluble Cu- protein, and the reduction of ferredoxin, a small FeS protein. Photosystem I is composed of a heterodimer of proteins that act as ligands for most of the electron carriers (Krauss *et al.*, 1993). The reaction centre is served by an antenna system that consists of about two hundred chlorophyll molecules (mainly chlorophyll *a*) and primary photochemistry is initiated by a chlorophyll *a* dimer, P700. In contrast to photosystem II, many of the antenna chlorophyll molecules in photosystem I are bound to the reaction centre proteins. Also, FeS centres serve as electron carriers in photosystem I and, so far as is known, photosystem I electron transfer is not coupled to proton translocation. Primary charge separation occurs between a primary donor, P700, a chlorophyll dimer, and a chlorophyll monomer (Ao).

Electron Transport

Electron transport from water to NADP+ requires three membrane bound protein complexes operating in series. Electrons are transferred between these large protein complexes by small mobile molecules (plastoquinone and plastocyanin in plants). Because these small molecules carry electrons (or hydrogen atoms) over relatively long distances, they play a unique role in photosynthetic energy conversion. Plastoquinone is hydrophobic, its movement is restricted to the hydrophobic core of the photosynthetic membrane. Plastoquinone operates by diffusing through the membrane until, due to random collisions, it becomes bound to a specific site on the photosystem II complex.

The photosystem II reaction centre reduces plastoquinone at the QB-site by adding two electrons and two protons creating PQH_2. The reduced plastoquinone molecule debinds from photosystem II and diffuses randomly in the photosynthetic membrane until it encounters a specific binding site on the cytochrome bf complex. The cytochrome bf complex is a membrane bound protein complex that contains four electron carriers, three cytochromes and an FeS centre. The crystal structure has been solved for cytochrome f from turnip (Martinez *et al.*, 1994) and the FeS centre from bovine heart mitochondria (Iwata *et al.*, 1996). In a complicated reaction sequence that is not fully understood, the cytochrome bf complex removes the electrons from reduced plastoquinone and facilitates the release of the protons into the inner aqueous space. The electrons are eventually transferred to the photosystem I reaction centre. The protons released into the inner aqueous space contribute to the proton chemical free energy across the membrane.

Electron transfer from the cytochrome bf complex to photosystem I is mediated by a small Cu-protein, plastocyanin (PC). Plastocyanin is water soluble and operates in the inner water space of the photosynthetic membrane. Electron transfer from photosystem I to NADP+ requires ferredoxin, a small FeS protein, and ferredoxin-NADP oxidoreductase, a peripheral flavoprotein that operates on the outer surface of the photosynthetic membrane. Ferredoxin and NADP+ are water soluble and are found in the outer aqueous phase.

The pathway of electrons is largely determined by the energetics of the reaction and the distance between the carriers. The electron affinity of the carriers is represented by their mid-point potentials, which show the free energy available for electron transfer reactions under equilibrium conditions. (It should be kept in mind that reaction conditions during photosynthesis are not in equilibrium.) Subsequent to primary charge separation, electron transport is energetically downhill (from a lower (more negative) to a higher (more positive) redox potential). It is the downhill flow of electrons that provides free energy for the creation of a proton chemical gradient.

Photosynthetic membranes effectively limit electron transport to two dimensions. For mobile electron carriers, limiting diffusion to two dimensions increases the number of random encounters (Whitmarsh, 1986). Furthermore, because plastocyanin is mobile, any one cytochrome bf complex can interact with a number of photosystem I complexes. The same is true for plastoquinone, which commonly operates at a stoichiometry of about six molecules per photosystem II complex.

Creation of a Proton Electrochemical Potential

Electron transport creates the proton electrochemical potential of the photosynthetic membrane by two types of reactions. (1) The release of protons during the oxidation of water by photosystem II and the translocation of protons from the outer aqueous phase to the inner aqueous phase by the coupled reactions of photosystem II and the cytochrome bf complex in reducing and oxidising plastoquinone on opposite sides of the membrane. This creates a concentration difference of protons across the membranes (DpH = pHin - pHout). (2) Primary charge separation at the reaction centre drives an electron across the photosynthetic membrane, which creates an electric potential across the membrane (DY = Yin - Yout). Together, these two forms of energy make up the proton electrochemical potential across the photosynthetic membrane (DmH+) which is related to the pH difference across the membrane and the electrical potential difference across the membrane by the following equation:

DmH+ = F DY - 2.3 RT DpH

where F is the Faraday constant, R is the gas constant, and T the temperature in Kelvin. Although the value of DY across the photosynthetic membrane in chloroplasts can be as large as 100 mV, under normal conditions the proton gradient dominates. For example, during photosynthesis the outer pH is typically near 8 and the inner pH is typically near 6, giving a pH difference of 2 across the membrane that is equivalent to 120 mV. Under these conditions the free energy for proton transfer from the inner to the outer aqueous phase is -12 kJ/ mol of protons.

Synthesis of ATP by the ATP Synthase Enzyme

The conversion of proton electrochemical energy into chemical free energy is accomplished by a single protein complex known as ATP synthase. This enzyme catalyses a phosphorylation reaction, which is the formation of ATP by the addition of inorganic phosphate (Pi) to ADP:

ADP-3 + Pi-2 + H+ → ATP-4 + H2O

The reaction is energetically uphill (DG = +32 kJ/mol) and is driven by proton transfer through the ATP synthase protein. The ATP Synthase complex is composed of two major subunits, CF_0 and CF_1. The CF_0 subunit spans the photosynthetic membrane and forms a proton channel through the membrane. The CF1 subunit is attached to the top of the CF_0 on the outside of the membrane and is located in the aqueous space. CF_1 is composed of several different protein subunits, referred to as a, b, g, d and e. The top portion of the CF_1 subunit is composed of three ab-dimers that contain the catalytic sites for ATP synthesis. A recent major breakthrough has been the elucidation of the structure of ATPase of beef heart mitochondria by Abrahams *et al.* (1994).

The molecular processes that couple proton transfer through the protein to the chemical addition of phosphate to ADP are poorly understood. It is known that phosphorylation can be driven by a pH gradient, a transmembrane electric field, or a combination of the two. Experiments indicate that three protons must pass through the ATP synthase complex for the synthesis of one molecule of ATP. However, the protons are not involved in the chemistry of adding phosphate to ADP. Paul Boyer and co-workers have proposed an alternating binding site mechanism for ATP synthesis (Boyer, 1993). One model based on their proposal is that there are three catalytic sites on each CF_1 that cycle among three different states. The states differ in their affinity for ADP, Pi and ATP. At any one time, each site is in a different state. This model is supported by the structure of ATPase elucidated by Abrahams *et al.* (1994).

Initially, one catalytic site on CF_1 binds one ADP and one inorganic phosphate molecule relatively loosely. Due to a conformational change of the protein, the site becomes a tight binding site, that stabilises ATP. Next, proton transfer induces an alteration in protein conformation that causes the site to release the ATP molecule into the aqueous phase. In this model, the energy from the proton electrochemical gradient is used to lower the affinity of the site for ATP, allowing its release to the water phase.

The three sites on CF_1 act cooperatively, i.e., the conformational states of the sites are linked. It has been proposed that protons affect the conformational change by driving the rotation of the top part (the three ab-dimers) of CF_1. Such a rotating model has recently been supported by recording of a rotation of the gamma subunit relative to the alpha-beta subunits by Sabbert *et al.* (1996). This revolving site mechanism would require rates as high as 100 revolutions per second. It is worth noting that flagella that propel some bacteria are driven by a proton pump and can rotate at 60 revolutions per second.

Synthesis of Carbohydrates

All plants and algae remove CO_2 from the environment and reduce it to carbohydrate by the Calvin cycle. The process is a sequence of biochemical reactions that reduce carbon and rearrange bonds to produce carbohydrate from CO_2 molecules. The first step is the addition of CO_2 to a five-carbon compound (ribulose 1,5-bisphosphate). The six-carbon compound is split, giving two molecules of a three-carbon compound (3-phosphoglycerate). This key reaction is catalysed by Rubisco, a large water soluble protein complex. The 3-dimensional structure has been determined by X-ray analysis for Rubisco isolated from tobacco (Schreuder *et al.* 1993) from a cyanobacterium (Synechococcus) (Newman and Gutteridge, 1993) and from a purple bacterium (Rhodospirillum rubrum) (Schneider *et al.* 1990). The carboxylation reaction is energetically downhill. The main energy input in the Calvin cycle is the phosphorylation by ATP and subsequent reduction by NADPH of the initial three-carbon compound forming a three-carbon sugar, triosephosphate.

Some of the triosephosphate is exported from the chloroplast and provides the building block for synthesising more complex molecules. In a process known as regeneration, the Calvin cycle uses some of the triosephosphate molecules to synthesise the energy rich ribulose 1,5-bisphosphate needed for the initial carboxylation reaction. This reaction requires the input of energy in the form of one ATP. Overall, thirteen enzymes are required to catalyse the reactions in the Calvin cycle.

The energy conversion efficiency of the Calvin cycle is approximately 90 per cent. The reactions do not involve energy transduction, but rather the rearrangement of chemical energy. Each molecule of CO_2 reduced to a sugar CH_2O requires 2 molecules of NADPH and 3 molecules of ATP.

Rubisco is a bifunctional enzyme that, in addition to binding CO_2 to ribulose bisphosphate, can also bind O_2. This oxygenation reaction produces the 3-phosphoglycerate that is used in the Calvin cycle and a two-carbon compound (2-phosphoglycolate) that is not useful for the plant. In response, a complicated set of reactions (known as photorespiration) are initiated that serve to recover reduced carbon and to remove phosphoglycolate. The Rubisco oxygenation reaction appears to serve no useful purpose for the plant. Some plants have evolved specialised structures and biochemical pathways that concentrate CO_2 near Rubisco. These pathways (C4 and CAM), serve to decrease the fraction of oxygenation reactions.

Photosynthetic Quantum Yield and Energy Conversion Efficiency

The theoretical minimum quantum requirement for photosynthesis is 8 quanta for each molecule of oxygen evolved (four quanta required by photosystem II and four by photosystem I). Measurements in algal cells and leaves under optimal conditions (e.g., low light) give quantum requirements of 8-10 photons per oxygen molecule released. These quantum yield measurements show that the quantum yields of photosystem II and photosystem I reaction centres under optimal conditions are near 100 per cent. These values can be used to calculate the theoretical energy conversion efficiency of photosynthesis (free energy stored as carbohydrate/light energy absorbed). If 8 red quanta are absorbed (8 mol of red photons are equivalent to 1,400 kJ) for each CO_2 molecule reduced (480 kJ/mol), the theoretical maximum energy efficiency for carbon reduction is 34 per cent.

Under optimal conditions, plants can achieve energy conversion efficiencies within 90 per cent of the theoretical maximum. However, under normal growing conditions the actual performance of the plant is far below these theoretical values. The factors that conspire to lower the quantum yield of photosynthesis include limitations imposed by biochemical reactions in the plant and environmental conditions that limit photosynthetic performance. One of the most efficient crop plants is sugar cane, which has to store up to 1 per cent of the incident visible radiation over a period of one year. However, most crops are less productive. The annual conversion efficiency of corn, wheat, rice, potatoes, and soybeans typically ranges from 0.1 per cent to 0.4 per cent (Odum, 1971).

Oxygenic Photosynthesis in Algae

Algae are photosynthetic eukaryotic organisms that, like plants, evolve O_2 and reduce CO_2. They represent a diverse group that include the dinoflagellates, the euglenoids, yellow-green algae, golden-brown algae, diatoms, red algae, brown algae, and green algae. The photosynthetic apparatus and biochemical pathways of carbon reduction of algae are similar to plants. Photosynthesis occurs in chloroplasts that contain photosystems II and I, the cytochrome bf complex, the Calvin cycle enzymes and pigment-protein complexes containing chlorophyll *a*, and other antenna pigments (e.g., chlorophyll *b* in green algae, chlorophyll *c* and fucaxanthol in brown algae and diatoms, and phycobilins in red algae). Green algae are thought to be the ancestral group from which land plants evolved. Algae are abundant and widespread on the earth, living mainly in fresh and sea water. Some algae live as single celled organisms, while others form multicellular organisms some of which can grow quite large, like kelp and seaweed. Phytoplankton in the ocean is made up of algae and oxygenic photosynthetic bacteria. Most photosynthesis in the ocean is due to phytoplankton, which is an imnortant source of food for marine life.

Oxygenic Photosynthesis in Bacteria

Cyanobacteria are photosynthetic prokaryotic organisms that evolve O_2 (Bryant, 1994). Fossil evidence indicates that cyanobacteria existed over 3 billion years ago and it is thought that they were the first oxygen evolving organisms on earth (Wilmotte, 1994). Cyanobacteria are presumed to have evolved in water in an atmosphere that lacked O_2. Initially, the O_2 released by cyanobacteria reacted with ferrous iron in the oceans and was not released into the atmosphere. Geological evidence indicates that the ferrous Fe was depleted around 2 billion years ago, and earth's atmosphere became aerobic. The release of O_2 into the atmosphere by cyanobacteria has had a profound affect on the evolution of life.

The photosynthetic apparatus of cyanobacteria is similar to that of chloroplasts. The main difference is in the antenna system. Cyanobacteria depend on chlorophyll *a* and specialised protein complexes (phycobilisomes) to gather light energy (Sidler, 1994). They do not contain chlorophyll *b*. As in chloroplasts, the chlorophyll *a* is located in membrane bound proteins. The phycobilisomes are bound to the outer side of the photosynthetic membrane and act to funnel exciton energy to the photosystem II reaction centre. They are composed of phycobiliproteins, protein subunits that contain covalently attached open ring structures known as bilins that are the light absorbing pigments. Primary photochemistry, electron transport, phosphorylation and carbon reduction occur much as they do in chloroplasts. Cyanobacteria have a simpler genetic system than plants and algae that enable them to be easily modified genetically. Because of this cyanobacteria have been used as a model to understand photosynthesis in plants. By genetically altering photosynthetic proteins, researchers can investigate the relationship between molecular structure and mechanism (Barry *et al.*, 1994).

Over the past three decades several types of oxygenic bacteria known as prochlorophytes (or oxychlorobacteria) have been discovered that have light harvesting protein complexes that contain chlorophyll *a* and *b*, but do not contain phycobilisomes (Palenik and Haselkorn 1992, Urbach *et al.*, 1992; Matthijs *et al.*, 1994). Because prochlorophytes have Chlorophyll *a/b* light harvesting proteins like chloroplasts, they are being investigated as models for plant photosynthesis.

Anoxygenic Photosynthesis

Anoxygenic photosynthetic bacteria differ from oxygenic organisms in that each species has only one type of reaction centre (Blankenship *et al.*, 1995). In some photosynthetic bacteria the reaction centre is similar to photosystem II and in others it is similar to photosystem I. However, neither of these two types of bacterial reaction centre is capable of extracting electrons from water, so they do not evolve O_2. Many species can only survive in environments that have a low concentration of O_2. To provide electrons for the reduction of CO_2, anoxygenic photosynthetic bacteria must oxidise inorganic or organic molecules available in their environment.

For example, the purple bacterium Rhodobacter sphaeroides can use succinate to reduce NAD+ by a membrane-linked reverse electron transfer that is driven by a transmembrane electrochemical potential. Although many photosynthetic bacteria depend on Rubisco and the Calvin cycle for the reduction of CO_2, some are able to fix atmospheric CO_2 by other biochemical pathways. Despite these differences, the general principles of energy transduction are the same in anoxygenic and oxygenic photosynthesis. Anoxygenic photosynthetic bacteria depend on bacteriochlorophyll, a family of molecules that are similar to the chlorophyll, that absorb strongly in the infra-red between 700 and 1000 nm. The antenna system consists of bacteriochlorophyll and carotenoids that serve a reaction centre where primary charge separation occurs. The electron carriers include quinone (e.g., ubiquinone, menaquinone) and the cytochrome bc complex, which is similar to the cytochrome bf complex of oxygenic photosynthetic apparatus. As in oxygenic photosynthesis, electron transfer is coupled to the generation of an electrochemical potential that drives phosphorylation by ATP synthase and the energy required for the reduction of CO_2 is provided by and ATP and NADH, a molecule similar to NADPH.

Purple Bacteria

There are two divisions of photosynthetic purple bacteria, the non-sulphur purple bacteria (e.g., Rhodobacter sphaeroides and Rhodospeudomonas viridis) and the sulphur purple bacteria (e.g., Chromatium vinosum) (Blankenship *et al.*, 1995). Non-sulphur purple bacteria typically use an organic electron donor, such as succinate or malate, but they can also use hydrogen gas. The sulphur bacteria use an inorganic sulphur compound, such as hydrogen sulphide as the electron donor. The only pathway for carbon fixation by purple bacteria is the Calvin cycle. Sulphur purple bacteria must fix CO_2 to live, whereas non-sulphur purple bacteria can grow aerobically in the dark by respiration on an organic carbon source.

The determination of the three-dimensional structures of the reaction centre of the non- sulphur purple bacteria, Rhodopseudomonas viridis and Rhodobacter sphaeroides, has provided an unprecedented opportunity to understand the structure and function of photosynthetic reaction centres

(Deisenhofer *et al.*, 1984, 1985; Feher *et al.*, 1989; Lancaster *et al.*, 1995). The reaction centre contains four bacteriochlorophyll and two bacteriopheophytin molecules. Two of the bacteriochlorophyll molecules form the primary donor (P870). At present, there is controversy over whether a bacteriochlorophyll molecule is an intermediate in electron transfer from the P870 to bacteriopheophytin. However, there is agreement that the remaining steps involve two quinone molecules (QA and QB) and that two turnovers of the reaction centre results in the release of reduced quinone (QH2) into the photosynthetic membrane.

Although there is a non-haem Fe between the two quinone molecules, there is convincing evidence that this Fe is not involved directly in transferring an electron from QA to QB. Because the primary donor (P870), bacteriopheophytin and quinone acceptors of the purple bacterial reaction centre are similar to the photosystem II reaction centre, the bacterial reaction centre is used as guide to understand the structure and function of photosystem II.

Light driven electron transfer is cyclic in Rhodobacter sphaeroides and other purple bacteria. The reaction centre produces reduced quinone, which is oxidised by the cytochrome bc complex. Electrons from the cytochrome bc complex are transferred to a soluble electron carrier, cytochrome c2, which reduces the oxidised primary donor P870+. The product of the light driven electron transfer reactions is ATP. The electrons for the reduction of carbon are extracted from an organic donor, such as succinate or malate or from hydrogen gas, but not by the reaction centre.

The energy needed to reduce NAD+ is provided by light driven cyclic electron transport in the form of ATP. The energy transformation pathway is complicated. Succinate is oxidised by a membrane bound enzyme (succinate dehydrogenase) that transfers the electrons to quinone, which is the source of electrons for the reduction of NAD+. However, electron transfer from reduced quinone to NAD+ is energetically uphill. By a mechanism that is poorly understood, a membrane bound enzyme is able to use energy stored in the proton electrochemical potential to drive electrons from reduced quinone to NAD+.

Green Sulphur Bacteria

Green sulphur bacteria (e.g., Chlorobium thiosulphatophilum and Chlorobium vibrioforme) can use sulphur compounds as the electron donor as well as organic hydrogen donors (Blankenship *et al.*, 1995). The reaction centre of green sulphur bacteria is similar to the photosystem I reaction centre of oxygenic organisms (Feiler and Hauska, 1995). The FeS centres in the reaction centre can reduce NAD+ (or NADP+) by ferredoxin and the ferredoxin-NAD(P)+ oxidoreductase enzyme, therefore green sulphur bacteria are not necessarily dependent on reverse electron flow for carbon reduction. The antenna system of the green sulphur bacteria is composed of bacteriochlorophyll and carotenoids and is contained in complexes known as a chlorosomes that are attached to the surface of the photosynthetic membrane. This antenna arrangement is similar to the phycobilisomes of cyanobacteria. Green sulphur bacteria can fix CO_2 without Rubisco. It has been proposed that they accomplish this by using the respiratory chain that normally oxidises carbon (known as the Krebs cycle), resulting in the release of CO_2. With the input of energy this process can be run in the reverse direction, resulting the uptake and reduction of CO_2.

Green Gliding Bacteria

Green gliding bacteria (e.g., Chloroflexus aurantiacus), also known as green filamentous bacteria, can grow photosynthetically under anaerobic conditions or in the dark by respiration under aerobic conditions. Like the green sulphur bacteria, green gliding bacteria harvest light using chlorosomes. The green gliding bacteria appear to have reaction centres similar to those of the purple bacteria, but there are several notable differences. For example, instead of two monomer bacteriochlorophyll molecules, C. aurantiacus has one bacteriochlorophyll and one bacteriopheophytin and the metal between the two quinones is Mn rather than Fe (Feick *et al.*, 1995). C. aurantiacus appears to fix CO_2 by a scheme that does not involve the Calvin cycle or the reverse Krebs cycle (Ivanovsky *et al.*, 1993).

Heliobacteria

Heliobacteria (e.g., Heliobacterium chlorum and Heliobacillus mobilis) are in the phylum Gram Positive Bacteria that are strict anaerobes. Although the heliobacterial reaction centre is similar to photosystem I in that it can reduce NAD+ (or NADP+), it contains a different type of chlorophyll known as bacteriochlorophyll g (Amesz, 1995).

Control of Intraprotein Electron Transfer

The three-dimensional structure of the reaction centre of Rhodopseudomonas viridis and Rhodobacter sphaeroides reveals the distances between the electron donors and acceptors (Deisenhofer *et al.* 1984,1985; Norris and van Brakel, 1986; Feher *et al.* 1989) and has had an important influence on biophysical and molecular genetics studies designed to identify the factors that control the rate of electron transfer within proteins. There is currently a controversy concerning the importance of specific amino acid composition of the protein on the rate of intraprotein electron transfer.

In part, the disagreement centres on whether the protein between the donor and acceptor molecules can be treated as a uniform material, or whether the specific amino acid composition of the protein significantly alters the rate. For example, it has been proposed that aromatic amino acids may provide a particular pathway that facilitates electron transfer between a donor and acceptor pair. This is the case in the photosystem II reaction centre, where a tyrosine residue on one of the reaction centre core proteins (precisely, Tyr 161 on the D1 protein) donates an electron to the primary donor chlorophyll, P680+. However, in other cases, replacement of an aromatic by another non-aromatic residue has resulted in relatively minor changes in the rate of electron transfer. L. Dutton and co-workers (Moser *et al.*, 1992) have analysed electron transfer reactions in biological and chemical systems in terms of electron tunnelling theory developed by R. Marcus and others (DeVault, 1984). Dutton and co-workers argue that protein provides a uniform electronic barrier to electron tunnelling and a uniform nuclear characteristic frequency. They suggest that the specific amino acid residues between an electron transfer pair is generally of less importance than the distance in determining the rate of pairwise electron transfer. In their view, protein controls the rate of electron transfer mainly through the distance between the donor and acceptor molecules, the free energy, and the reorganisation energy of the reaction.

The importance of distance is demonstrated by electron transfer data from biological and synthetic systems showing that the dependence of the electron transport rate on the edge to edge distance is exponential over 12-orders of magnitude when the free energy is optimised (Moser *et al.*, 1992). The extent to which this view is generally applicable for intraprotein transfer remains to be established (Williams, 1992).

One of the challenges in understanding pairwise electron transfer rates from first principles is illustrated by the reaction centres of Rhodopsuedobacter sphaeroides in which the redox components are arranged along twofold axis of symmetry that extends from the primary donor (P870) to the non- haem Fe. Despite the fact that the reaction centre presents two spatially similar pathways for electron transfer from P870 to quinone, nearly all electrons are transferred down the right-arm of the reaction centre. The same is true for the reaction centre of Rhodopseudomonas viridis, in which it is estimated that electron transfer down the left-arm is less than 1:100 (Kellogg *et al.*, 1989).

The challenge to theorists is to explain the surprisingly high probability that electron flow goes down the right-arm. Since the distances are similar, it has been suggested that electron transfer down the left-arm is less probable due to an endothermic free energy change (Parson *et al.*, 1990) or to an unfavourable rearrangement energy for the reaction (Moser *et al.*, 1992).

Global Photosynthesis and the Atmosphere

The amount of CO_2 removed from the atmosphere each year by oxygenic photosynthetic organisms is massive. It is estimated that photosynthetic organisms remove 100 x 1015 grams of carbon (C)/year (Houghton and Woodwell, 1990). This is equivalent to 4 x 1018 kJ of free energy stored in reduced carbon, which is roughly 0.1 per cent of the incident visible radiant energy incident on the earth/year. Each year the photosynthetically reduced carbon is oxidised, either by living organisms for their survival, or by combustion. The result is that more CO_2 is released into the atmosphere from the biota than is taken up by photosynthesis. The amount of carbon released by the biota is estimated to be 1-2 x 1015 grams of carbon/year. Added to this is carbon released by the burning of fossil fuels, which amounts to 5 x 1015 grams of carbon/year. The oceans mitigate this increase by acting as a sink for atmospheric CO_2. It is estimated that the oceans remove about 2 x 1015 grams of carbon/year from the atmosphere. This carbon is eventually stored on the ocean floor. Although these estimates of sources and sinks are uncertain, the net global CO_2 concentration is increasing. Direct measurements show that each year the atmospheric carbon content is currently increasing by about 3 x 1015 grams. Over the past two hundred years, CO_2 in the atmosphere has increased from about 280 parts per million (ppm) to its current level of 360 ppm. The consequences of this rapid change in our atmosphere are unknown. Because CO_2 acts as a greenhouse gas, some climate models predict that the temperature of the earth's atmosphere may increase by 2-8°C. Such a large temperature increase would lead to significant changes in rainfall patterns. Little is known about the impact of such drastic atmospheric and climatic changes on plant communities and crops. Current research is directed at understanding the interaction between global climate change and photosynthetic organisms.

Biological Energy Production

Biological Energy Production

Energy and Environmental Issues: The widespread use of fossil fuels, has brought numerous benefits to industrialised societies. Large amounts of agricultural, domestic and industrial wastes generated in these countries as a result of development, have potentially detrimental effects both on the environment and on human health. Itai-itai and Minamata diseases in Japan, are just two examples of the effects of air and water pollution on human health. The importance of protecting the environment and restoring environmental damage cannot be overemphasised.

In recent years, environmental pollution has become a global problem. Internationalisation of industrial and social activities has given rise to problems such as global warming, desertification, and acid deposition. These global problems are rooted in the materially-rich lifestyles which are supported by abundant and wasteful use of fossil fuels in industrialised countries. Rapidly increasing industrial activities in China, India, and in other developing countries implicates that these countries will inevitably contribute to deterioration of the global environment and to destruction of the global ecosystem. Lifestyle changes, and changes in our key industrial systems are required in order to minimise the impact of environmental pollution. The recycling of materials, and thus minimising the generation of waste, is a basic concept which must be implemented in order to meet the new demands of sustainable development in both industrialised and developing countries.

Mechanisms for implementing this concept and for establishing environmentally compatible technologies which support the future "recycling" world are required. Systems, which utilise energies produced from biomass are typical examples of energy recycling systems. Biotechnology is one of the future-oriented technologies, and one that will play a major role in the exploitation of biomass energy. All biomass (plant, animal and microbial), originates through CO_2 fixation by photosynthesis. Biomass utilisation is consequently included in the global carbon cycle of the biosphere. Biomass energy in developing countries, originates from fuelwood, animal wastes, and agricultural residues, and is primarily utilised for activities which are essential to survival, such as cooking and obtaining water. Improvements in the living standards in these countries will result in the non-essential use of energy. Development of technologies that efficiently produce biomass, and convert it to more convenient forms of energy is therefore very important.

Photosynthesis and Biomass

Photosynthetic Efficiency: Photosynthesis can be simply represented by the equation:

$$6CO_2 + 12H_2O \text{ (+light energy)} \longrightarrow C_6H_{12}O_6 + 6O_2 + 6H_2O$$

Approximately 114 kilocalories of free energy are stored

in plant biomass for every mole of CO_2 fixed during photosynthesis. Solar radiation striking the earth on an annual basis is equivalent to 178,000 terawatts, i.e. 15,000 times that of current global energy consumption. Although photosynthetic energy capture is estimated to be ten times that of global annual energy consumption, only a small part of this solar radiation is used for photosynthesis. Approximately two thirds of the net global photosynthetic productivity worldwide is of terrestrial origin, while the remainder is produced mainly by phytoplankton (microalgae) in the oceans which cover approximately 70 per cent of the total surface area of the earth. Since biomass originates from plant and algal photosynthesis, both terrestrial plants and microalgae are appropriate targets for scientific studies relevant to biomass energy production.

Any analysis of biomass energy production must consider the potential efficiency of the processes involved. Although photosynthesis is fundamental to the conversion of solar radiation into stored biomass energy, its theoretically achievable efficiency is limited both by the limited wavelength range applicable to photosynthesis, and the quantum requirements of the photosynthetic process. Only light within the wavelength range of 400 to 700 nm (photosynthetically active radiation, PAR) can be utilised by plants, effectively allowing only 45 per cent of total solar energy to be utilised for photosynthesis. Furthermore, fixation of one CO_2 molecule during photosynthesis, necessitates a quantum requirement of ten (or more), which results in a maximum utilisation of only 25 per cent of the PAR absorbed by the photosynthetic system. On the basis of these limitations, the theoretical maximum efficiency of solar energy conversion is approximately 11 per cent. In practice, however, the magnitude of photosynthetic efficiency observed in the field, is further decreased by factors such as poor absorption of sunlight due to its reflection, respiration requirements of photosynthesis and the need for optimal solar radiation levels. The net result being an overall photosynthetic efficiency of between 3 and 6 per cent of total solar radiation.

Biomass Wastes and their Conversion

Wastes and residues currently constitute a large source of biomass (1). These include solid and liquid municipal wastes, manure, lumber and pulp mill wastes, and forest and agricultural residues. With the exception of feedstocks of low water content, most of this biomass cannot be directly utilised, and must undergo some form of transformation, prior to being utilised as a fuel. Biological processes for the conversion of biomass to fuels include ethanol fermentation by yeast or bacteria, and methane production by microbial consortia under anaerobic conditions.

Wood wastes in the paper and pulp industries and bagasse from the sugar-cane industry are examples of biomass likely to accumulate at a single site. The cellulosic nature of these biomass materials, necessitates their hydrolysis to glucose, prior to ethanol fermentation. The net energy balance for the processes involved can, however, be problematic in that energy requirements for cellulose hydrolysis and distillation, must be lower than the energy in the output ethanol.

Unlike ethanol fermentation, anaerobic digestion for methane production, utilises organic materials containing carbohydrates, lipids, and proteins. Many species of microbes work cooperatively in an anaerobic digester, in which these polymeric materials (i.e. carbohydrates, proteins and lipids) are first decomposed to organic acids, and then to hydrogen and carbon dioxide, from which methane is synthesised by methanogens. A variety of raw materials which include agricultural wastes, municipal solid wastes, market garbage, and waste water from food and fermentation industries, are applicable as substrates for this process. Waste products derived from animal husbandry are applicable to anaerobic digestion, with the added bonus of solving the environmental issues of unpleasant odours and eutrophication. Although small-scale digesters are popularly used at both the farm and village levels, large-scale operations are still in need of considerable technical improvement and cost reduction, and thus require both microbial and engineering studies.

Methane and ethanol can also be produced from cultured microalgal biomass through anaerobic digestion and microbial fermentation processes, respectively. The economics of fuel production from microalgal biomass is however largely dependent on a microalgal CO_2 fixation step similar to that required for the production of H_2 and algal oils.

Fuel Production via Microalgal CO_2 Fixation

One of the most serious environmental problems today is that of global warming, caused primarily by the heavy use of fossil fuels. In Japan, large amounts of CO_2 are released into the atmosphere from electric power plants and industry. The CO_2 generated by these large point sources could potentially be recovered with relative ease through the use of an established technology such as chemical absorption. The enormity of the amounts of potentially recoverable CO_2 would however necessitate the development of technologies for sequestering or, more favourably, utilising this CO_2.

Photosynthetic microalgae are potential candidates for utilising excessive amounts of CO_2, since when cultivated these organisms are capable of fixing CO_2 to produce energy and chemical compounds upon exposure to sunlight. The derivation of energy from algal biomass is an attractive concept in that unlike fossil fuels, algal biomass is rather uniformly distributed over much of the earth's surface, and its utilisation would make no net contribution to increasing atmospheric CO_2 levels. Although algal biomass is regarded as a low-grade energy source owing to its high moisture content, through biological processes, it may be converted to modem gaseous and liquid fuels such as hydrogen, methane, ethanol, and oils.

Hydrogen is regarded as a potential energy source of the future, since it is easily converted to electricity and bums cleanly. Hydrogen is currently produced by fossil fuel-based processes which emit large amounts of CO_2, and relatively smaller amounts of other air pollutants such as sulphur dioxide and nitrogen oxides. Biological H_2 production has thus recently received renewed attention owing to urban air pollution and global warming concerns. Biological hydrogen production methodologies incorporating artificial reconstitution systems with chloroplast, ferredoxin, and hydrogenase; a heterocystous cyanobacterial system with oxygen scavengers; and an algal system in a day and night cycle, have been studied in Japan. From an engineering point of view, however, bacterial fermentation mechanisms for hydrogen production under either dark or light conditions is currently of importance in terms of environmental issues and the utilisation of organic wastes such as waste effluent of the food and fermentation industries, pre-treated sewage sludge, and market garbage.

The use of microalgae as sources of liquid fuels is an attractive proposition from the point of view that microalgae are photosynthetic renewable resources, are of a high lipid content, have faster growth rates than plant cells, and are capable of growth in saline waters which are unsuitable for agriculture. While the lipid content of microalgae, on a dry cellular weight basis varies between 20 and 40 per cent, lipid contents as high as 85 per cent have been reported for certain microalgal strains. *Botryococcus braunii*, is a unique microalgal strain, having a long-chain hydrocarbon content of between 30 and 40 per cent (dry weight basis), which is directly extractable to yield crude oil substitutes. Both physical and chemical processes are applicable in the production of liquid fuels from algal strains of high lipid content. These processes include direct lipid extraction in the production of diesel-oil substitutes, transesterification in the formation of ester fuels, and hydrogenation in the production of hydrocarbons. Oily substances are also produced via liquefaction of microalgal biomass through thermochemical reactions under conditions of high pressure and temperature.

General Problems

Among the biomass conversion processes discussed thus far, methane and ethanol production from various wastes is economically feasible within the restraints of scale and location.

Although biological processes for the production of gaseous and liquid fuels have been well demonstrated with cultured microalgal biomass, these processes must still be integrated into a system capable of meeting basic requirements for overall efficiency of converting solar energy into biofuels. Furthermore, a model system must at least in principle, be capable of easy scale-up and not be limited by either engineering or economic factors. Under the current petroleum economy, prospects for the use of H_2 or oils produced by biological processes seem remote. However, future requirements for a "clean environment" necessitate fundamental research into microbial and algal physiology and genetics, together with basic engineering research on converters and total systems.

In its effort to address global warming and other environmental problems, the Research Institute of Innovative Technology for the Earth (RITE), engaged in investigating projects, including the Project for Biological CO_2 Fixation and Utilisation, and the Project for Biological Production of Hydrogen through the use of environmentally acceptable technologies was established by the Japanese Government in 1990. These projects employ researchers from both national institutes and private companies, who are engaged in the research and development of technologies for microbiological CO_2 fixation and the production of hydrogen. At the same time, research efforts at the university level include biochemical and genetic studies on CO_2-fixing and H_2-producing enzymes, and the application of these enzymes to the development of more efficient energy systems.

Energy Conversion by Photosynthetic Organisms

Energy demands are nevertheless expected to continue increasing with global improvements in living standards. There is however a limit to the amount of energy consumption that can be tolerated by nature. Excessively large amounts of energy produced by nuclear fusion for example, have the potential of altering terrestrial climates. Renewable sources of energy which do not alter global energy levels are therefore desirable.

Photosynthetic Capture of Solar Energy

Solar Energy: Approximately 5.7 x 10^{24} J of solar energy are irradiated to the earth's surface on an annual basis. Plants and photosynthetic organisms utilise this solar energy in fixing large amounts of CO_2 (2x10^{11} t = 3x10^{21} J/year), while amounts consumed by human beings are relatively small, (3 x 10^{20} J/year) (1), representing only 10 per cent of the energy converted during photosynthesis.

Although large amounts of solar energy are irradiated to the earth's surface, the effective energy concentration (energy/unit area) of solar energy at any one point on the earth's surface is small—only about 1 kW/m^2 at most, even at noon. Such low effective energy concentrations, limit the use of solar energy as a primary energy source, and elevate the costs associated with its accumulation and transmission.

Technologies for the utilisation of low-density energy sources must be developed in order to facilitate the use of solar energy. Solar energy conversions through the use of photosynthetic microorganisms do not incorporate the use of complex systems or large quantities of factory manufactured products, and indeed have relatively minimal investment and resource requirements. Additionally, these technologies are largely dependent on the use of renewable resources, thereby generating minimal amounts of waste. Recent advances in biotechnology have made possible studies on the utilisation of biological processes such as photosynthesis for energy production.

Why is biotechnology now applied to energy technology?

The use of natural energy involves the control of entropy. Prior to the industrial revolution, wood served as a major energy source. However, the industrial revolution gave rise to the widespread use of both coal and petroleum as energy sources. Since coal and petroleum were often produced at sites far from their consumption points, recuperation of their production costs involved mass production, thus precipitating the formation of an integrated industrial society. This integrated system soon reached its limits, making it necessary to consider discrete social systems that utilise delocalised abundant energy sources.

Integration is necessary in any system where energy, goods, and information are dispersed. Integration of systems involves the use of entropy-reducing processes. Application of mechanical methods to the integration of systems incorporates large amounts of energy expenditure. Existing problems cannot therefore be resolved without modification of existing social systems. Consequently, energy production technology must be reformed. Microorganisms have the ability to reduce entropy through energy utilisation, and can potentially simplify the conversion and accumulation of solar energy and energy utilisation over large areas.

Following the industrial revolution, energy-releasing techniques were developed. The development of techniques which control entropy are essential for human survival. To this end, there is a need for the development of industrial technology which makes use of biological principles in a sophisticated manner.

Photosynthesis Mechanisms

Biological energy conversions can be categorised into two groups: i) photosynthesis (the process whereby solar energy is fixed to yield energy useful to organisms and industry), and ii) biomass conversion (the product of photosynthesis) into energy. Photosynthesis occurs in plants, algae and photosynthetic bacteria, while biomass conversion reactions often occur in non-photosynthetic microorganisms.

Plant Photosynthesis

Photosynthesis is often regarded as a CO_2 anabolic reaction, whereby glucose is formed from CO_2 and water. CO_2 anabolism is an energy-consuming reaction in that it utilises chemical energy produced by photosynthesis. In its narrowest sense, photosynthesis can be regarded as a process whereby energy is supplied for CO_2 anabolism. In a broader sense, photosynthesis, including CO_2 anabolism, can be divided into several steps: i) photoelectric charge isolation using photon energy (conversion to electrical energy), ii) fixation of electrical energy in the form of chemical energy (ATP synthesis), and iii) chemical reactions involving ATP (fixation of CO_2, and hydrogen production).

The supply of energy for CO_2 anabolism is common to all photosynthetic organisms which exhibit photosynthesis. Energy conversion, ATP synthesis and the production of both CO_2 and hydrogen on the other hand, are not unique to photosynthetic organisms, but occur in all types of microorganisms, and are in fact similar to the respiratory processes which occur in mitochondria of higher organisms. Two types of photosynthesis are distinguishable on the basis of source of the electrons used as energy carriers. In plants such as green algae, and cyanobacteria (blue-green algae), water is the electron source, while in photosynthetic bacteria, organic or sulphur compounds provide electron sources. Two Photosystem II water molecules are initially decomposed by four incident photons, to yield one oxygen molecule and four excited electrons. Excited electron energy is subsequently utilised in ATP synthesis. Unlike in the case of ordinary chemical reactions, ATP synthesis cannot be stoichiometrically analysed.

The ratio of excited photons to ATP produced is still a somewhat debatable issue. Although it has generally been thought that two photons give rise to the formation of two ATP molecules, some researchers claim that three photons are involved (3). Furthermore, other researchers have suggested a loose coupling between proton transport and ATP synthesis:

$$4 \text{ photons} + 2H_2O + 2(\text{or more})ADP = 2(\text{or more})ATP + 4H^+ + 4e^- + O_2$$

Subsequent to their energy release in ATP production, photosystem II electrons are transported to photosystem I, where they are again excited to a higher energy level, allowing them to be utilised for NADP reduction. NADP serves both as an electron carrier and an oxidising and reducing agent *in vivo*. Two photons are utilised per molecule of NADP reduced:

$$4 \text{ photons} + 4e^- + 2NADP + 4H^+ = 2NADPH.$$

Photosystem I may also be involved in ATP synthesis. In cases where it is involved, excited photosystem I electrons are recycled:

$$4 \text{ photons} + 2(\text{or more})ADP = 2(\text{or more})ATP.$$

Fixation of one molecule of CO_2, involves the following reaction:

$$CO_2 + 3ATP + 2NADPH = CH_2O + 3ADP + 2NADP.$$

If two ATP molecules are obtained through Photosystem II excitation (Eq. 2-1), the net reaction, following equations 2-1 through 2-4 is:

$$CO_2 + 10 \text{ photons} + H_2O = CH_2O + 1/2O_2.$$

Figure: Schematic representation of mechanisms involved in plant photosynthesis

Experimental data indicates that between 8 and 12 photons are required for fixation of one molecule of CO_2. Since the energy equivalent of one photon (700 nm) is approximately 170 kJ/E, and the change in free energy during the fixation of CO_2 is approximately 450 U/mol, the energy efficiency of this process for monochromatic light of a wavelength of 700 nm is estimated to be approximately 21-33 per cent. However, owing to the quantum nature of photosynthetic reactions, energy efficiency decreases if light of shorter wavelengths (i.e. higher quantum energy) is used. Additionally, energy losses, energy requirements for plant growth, and the distribution of solar energy wavelengths need to be considered. Plant photosynthesis takes place only in the presence of visible light (400-700 nm). However, solar light contains both visible and infrared components. Since visible light accounts for about 45 per cent of all solar energy, the maximum achievable energy efficiency for CO_2 fixation using solar radiation is approximately 13 per cent.

Bacterial Photosynthesis

Bacterial photosynthesis is thought to be a relatively old form of photosynthesis. It incorporates the use of either organic or sulphur compounds as electron donors in photosystem I. Unlike in the case of plant photosynthesis, cyclic photophosphorylation takes place in bacterial photosynthesis, i.e. electrons are repeatedly excited in a cyclic manner, with ATP being generated in each cycle. Photosynthetic bacteria are also capable of reducing electron carriers such as NAD, via a linear reaction similar to the electron transmission which occurs during plant photosynthesis.

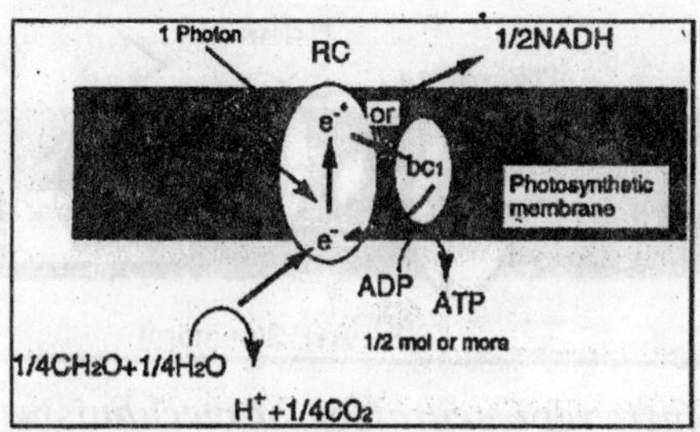

Figure: Schematic representation of mechanisms involved in bacterial photosynthesis

CO_2-fixing reactions do not produce energy during bacterial photosynthesis (i.e. equimolar amounts of organic compounds are produced through decomposition of organic compounds), except when sulphur compounds serve as electron carriers.

2 photons + 1 (or more)ADP = 1 (or more)ATP.

Electrons are donated as follows:

$CH_2O + H_2O = 4e^- + 4H^+ + CO_2$.

The structure of the photosynthetic reaction centre (RC), involved in the early steps of photosynthesis, has been elucidated for certain photosynthetic bacteria. Such chlorophyll- containing bacteria which include *Rhodopseudomonas viridis* and *Rhodobacter sphaeroides,* show similarities with respect to the arrangement of chlorophyll, and the three-dimensional structures of major portions of the proteins possessing that pigment.

Such structural similarities between photosynthetic bacteria, seem to suggest the acquisition of an optimal structure by these bacteria, over a long evolutionary period.

Pigments such as bacteriochlorophyll are also present within the RC. Photoelectric charge isolation takes place within dimers of these bacteriochlorphyll pigments, resulting in the release of high-energy electrons, *via* the action of bacteriochlorophyll monomers such as bacteriopheophytin, quinone A, and quinone B.

These high-energy electrons are subsequently conjugated with proton transportation in the cytochrome b/c1 complex.

A noteworthy feature of the RC function is that photon involvement in photoelectric charge isolation, resembles that which occurs in photo-semiconductors.

These RC centres can thus be regarded as molecular elements produced by nature. The fact that photoelectric charge isolation is observed in protein molecules will greatly influence future research relevant to molecular elements and solar batteries.

Hydrogen Production through Solar Energy Conversions

Although large amounts of solar energy are irradiated to the earth's surface, the actual concentration (energy/unit area) of solar energy at the earth's surface is relatively low. The accumulation of solar energy for practical use, therefore necessitates collection of solar energy over large areas. This could potentially be a costly venture, particularly with respect to the use of solar batteries for solar energy concentration. Photosynthetic bacteria however have the potential to eliminate requirements for such large batteries for solar collection.

Hydrogen appears to be the most useful form of solar-converted energy, in that it can be easily substituted for petroleum-based fuels. The use of hydrogen is also advantageous in that when burnt it yields only water, and hence does not contribute to environmental pollution.

Cyanobacterial Hydrogen Production (Plant-type Photosynthesis)

Cyanobacteria (blue-green algae) and photosynthetic bacteria are representative photosynthetic hydrogen producers. *In vivo* hydrogen production by these microorganisms is catalysed by either nitrogenase or hydrogenase enzymes. Nitrogenases require ATP for catalytic action:

$$2H^+ + 2e^{-*}(Fd) + 4ATP = H_2 + 4ADP.$$

Hydrogenases which catalyse hydrogen production in organisms such as *Clostridium*, on the other hand, do not require ATP for their activity. However, since hydrogenase activity involves some hydrogen uptake, the efficiency of hydrogen production through hydrogenase activity is low. Relatively little hydrogen is produced by these organisms in the absence of strong electron donors such as ferredoxin (Fd):

$$2H^+ + e^{-*} = H_2$$

(e^{-*}: high-energy electrons in NADH, Fd_{red}, etc.).

Where hydrogenase enzymes are used for hydrogen

production and the reducing potential, yielded by photosynthetic reactions is completely used for reducing hydrogen ions, two photons can theoretically produce one hydrogen molecule:

$$2 \text{ photons} + H_2O + 1(\text{or more})ADP = 1 (\text{or more})ATP + 2H^+ + 2e^- + 1/2O_2$$

$$2 \text{ photons} + 2e^- = 2e^{-*}$$

$$2H^+ + 2e^{-*} = H_2 \text{ (by hydrogenase).}$$

Hydrogenase-mediated reactions in cyanobacteria are not usually biased in favour of hydrogen production. Instead, such reactions are likely to cause hydrogen uptake and are consequently impractical. It is highly likely that the high-energy electrons, yielded as a result of *Clostridium)* or by the use of these organic substances as electron donors. The efficiency of the latter reaction can be determined by multiplying the efficiency of CO_2; anabolism by the efficiency of hydrogen production in photosynthetic bacteria.

Although there are few reports of precisely determined hydrogen production efficiencies of cyanobacteria, Miyamoto et al. determined that outdoor solar incubation for a period exceeding one month in California, resulted in an average energy conversion efficiency (energy yielded by combustion of produced hydrogen/incidence solar energy) of 0.2 per cent (6).

Bacterial Hydrogen Production (Bacterial-type Photosynthesis)

Upon exposure of ammonia-free media containing photosynthetic bacteria to light, nitrogenase activity is induced, resulting in hydrogen production. Organic substances such as lactic acid serve as electron donors in photosynthetic bacteria. In such reactions, "G is positive, indicating that the use of solar energy allows photosynthetic bacteria to produce hydrogen through complete decomposition of organic substances. Anaerobes such as *Clostridium* also produce hydrogen, but are incapable of completely utilising energy or decomposing organic substances:

$$C_3H_6O_3 + 3H_2O = 12H^+ + 3CO_2 + 12e^- = 6H_2 + 3CO_2 \text{ ''}G = 51.2 \text{ kJ}$$

$$C_6H_{12}O_6 + 2H_2O = 4H_2 + 2CH_3COOH + 2CO_2$$

$$\text{''}G = -184.2 \text{ kJ}.$$

Photosynthetic bacteria are capable of completely decomposing organic substances. Studies on hydrogen production through the exposure of organic wastes (waste fluids from food factories, pulp factories, etc.) to photosynthetic bacteria and light, have been conducted in a number of countries. Research into hydrogen production using a combination of photosynthetic bacteria and anaerobes has also been conducted.

A key factor in determining the commercial applicability of hydrogen production processes, is the rate at which hydrogen is produced. Bacteria have been widely investigated for their rates of hydrogen production. To date, *R. sphaeroides* has been identified as the bacterium having the highest hydrogen-producing rate (260 ml/mg/h), with a photo-energy conversion efficiency (energy yielded by combustion of produced hydrogen/incident solar energy) of 7 per cent, determined using a solar simulator.

Further strain development will potentially elevate the energy conversion efficiency of photosynthetic bacteria to levels comparable to those of solar batteries.

Organic substances are utilised as electron donors by photosynthetic bacteria. The energy required for extracting electrons from these molecules is much lower than that required for the hydrolysis of water. Photo-energy is more often used for nitrogenase activation, i.e. ATP reproduction, than for the decomposition of organic substances.

The number of photons required for ATP reproduction by

photosynthetic bacteria has not been theoretically elucidated, though an experimental number of 1.5 has been determined. A value of 1 has also been suggested.

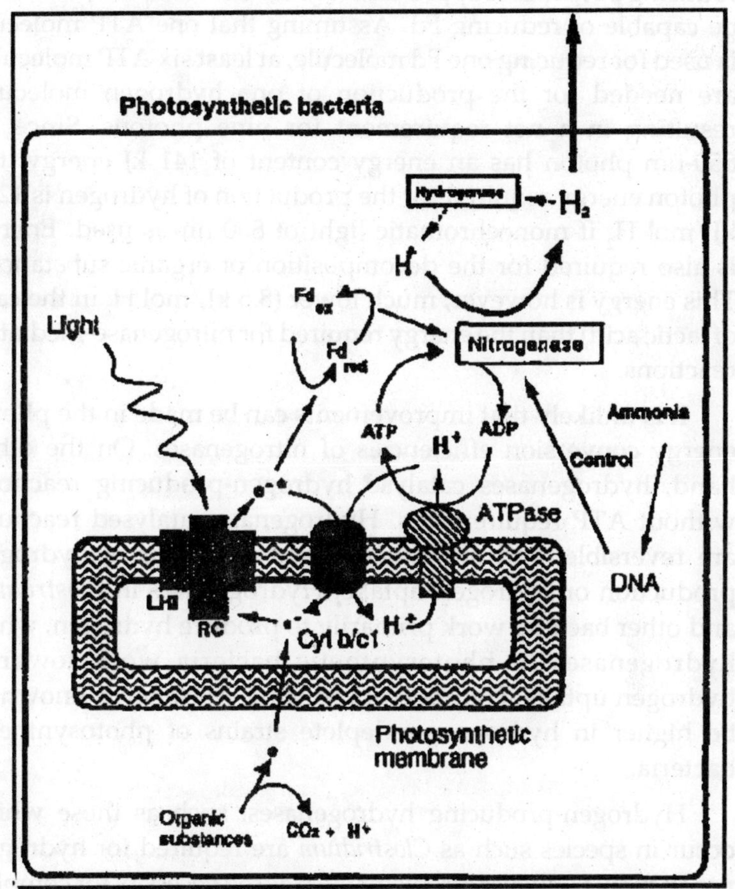

Figure: Mechanism for hydrogen production by photosynthetic bacteria

In addition to ATP, nitrogenase-mediated reactions, require Fd as an electron carrier. Reaction mechanisms involved have not however been completely elucidated. The reducing potential, created in the RC of purple, non-sulphur photosynthetic bacteria such as *R. sphaeroides* and *R. capsulatus* is approximately 100 mV at most. Such bacteria would be incapable of directly reducing Fd, which has an oxidising and reducing potential of approximately 400 mV. ATP may however be capable of reducing Fd. Assuming that one ATP molecule is used for reducing one Fd molecule, at least six ATP molecules are needed for the production of one hydrogen molecule, resulting in a net requirement for nine photons. Since an 850-nm photon has an energy content of 141 kJ energy, the photon energy required for the production of hydrogen is 1269 kJ/mol H_2 if monochromatic light of 850 nm is used. Energy is also required for the decomposition of organic substances. This energy is however, much lower (8.5 kJ/mol H_2 in the case of lactic acid) than the energy required for nitrogenase-mediated reactions.

It is unlikely that improvements can be made in the photo-energy conversion efficiencies of nitrogenases. On the other hand, hydrogenases catalyse hydrogen-producing reactions without ATP requirements. Hydrogenase-catalysed reactions are reversible, and are either biased in favour of hydrogen production or hydrogen uptake. Hydrogenases in *Clostridium* and other bacteria work primarily to produce hydrogen, while hydrogenases in photosynthetic bacteria work towards hydrogen uptake. Hydrogen-producing efficiency is known to be higher in hydrogenase-deplete strains of photosynthetic bacteria.

Hydrogen-producing hydrogenases, such as those which occur in species such as *Clostridium* are required for hydrogen production. However, problems such as the necessity for complex control are encountered with the use of hydrogenases. Technologies for the incorporation of hydrogenases into reaction systems must therefore be developed.

Use of Photosynthesised Proteins in Photoelectric Conversion Elements

As stated above, RC is capable of converting photo-energy into electrical energy. It can therefore be utilised in either photoelectric converters or photo-semiconductors. Solar batteries containing chromatophore membranes or an RC are now available. The RC is a relatively stable protein which retains its photoelectric conversion function even in the dry state. Unlike ordinary enzymes, the RC neither binds to the substrate nor undergoes chemical reactions, except for its binding to, or dissociation from quinones. Upon exposure to photons within membranes, the RC only releases electrons and has no mobile portions. Unlike enzymes, the RC does not need to vibrate within water molecules. Using the Langmuir-Blodgett method, it is possible to prepare dried RC photoelectric devices.

Use of Immobilised Yeast Cells in Alcohol Fermentations

Alcohol fermentation is a well established technology which has long been practised throughout the world. A fermentation process incorporating the flash method which is seen as a means of improving alcohol productivity using yeasts with poor alcohol resistance is also discussed here.

Preparation of Immobilised Yeast Cells

Of the various methods used for continuous fermentation, that employing the use of immobilised yeast cells was selected. Methods of yeast immobilisation evaluated included collagen casting, acetic acid-lactic acid cellulose microencapsulation, chitosan/glutar-aldehyde moulding, carrageenan-entrapping method, and calcium alginate gel-entrapping. Of these, the calcium alginate gel-entrapping method was preferred because of its high enzymatic activity, simple manner of preparation, and stability. Preparation of a uniform calcium alginate gel, necessitated maintaining the viscosity of the mixture of calcium alginate and yeast cells between 1000 and 2000 cps.

Continuous Plant Operation using Immobilised Yeast Cells

The spherical gel method was employed for the preparation of calcium alginate gels since this method does not require the use of specialised equipment. Spherical gels are readily obtained by adding sodium alginate solution to calcium chloride solution using a nozzle. No special granulation apparatus was used when the equipment was assembled, but a gel-dropping nozzle was provided at the top of the fermentor. The fermentor was filled with a calcium chloride solution prior to fermentation, and sodium alginate solution was added dropwise to form granules. The culture medium was then supplied to the fermentor to initiate the fermentation. This procedure was found to simplify the gel preparation process. Continuous fermentation tests were conducted using two 1-kL fermentors arranged in series. Continuous production of alcohol was carried out stably for 4000 hours while maintaining high enzymatic activity.

Fermentation Processes used in Ethanol Production

Alcohol productivity has been greatly improved through the development of bioreactors.

Flash Fermentation Using Immobilised Yeast Cells

Alcohol inhibition of yeasts, does not pose any problems in alcohol fermentation by the batch method, since yeasts are used only once in batch processes. In such processes, yeast damage due to alcoholic inhibition is of relatively little significance. However, where immobilised continuous methods of alcohol production are used, stable yeast activity needs to be maintained, and the alcohol concentration in the fermentor must be relatively low. Several methods of continuously removing alcohol from fermentors as a means of avoiding alcoholic inhibition of yeasts have been proposed. The fermentor used in our study was equipped with a distillation column (operable either under atmospheric or reduced pressure). Circulation of the fermentation broth through this distillation column, enabled the removal of alcohol, and sugar thereby facilitating the maintenance of low alcohol concentrations within the fermentor. Alcoholic inhibition of yeast was reduced while the fermentation rate increased. In addition, this method enabled raw material to be used at a high concentration, reducing the volume of waste liquor. However, the use of raw material at a high concentration necessitates a relatively longer retention time within the fermentor, thus increasing the accumulation of by-products, which inevitably reduces the fermentation rate due to by-product inhibition.

Alcohol Production Using an Integrated Pilot Plant

Various aspects of alcohol fermentation from cellulosic biomass have been discussed thus far. A number of problems still remain to be resolved prior to industrial-scale production of fuel alcohol from cellulosic biomass. As mentioned in the introduction to this chapter, subsequent to our studies on basic and elemental techniques, an integrated pilot plant for the production of alcohol from biomass, was constructed in our laboratory, in order to demonstrate individual processes.

Construction of the plant commenced in 1983, and continued in a step-wise manner for 5 years. The final plant was capable of treating 720 kg of raw material per day with the production of 150 to 200 litres of dehydrated fuel alcohol.

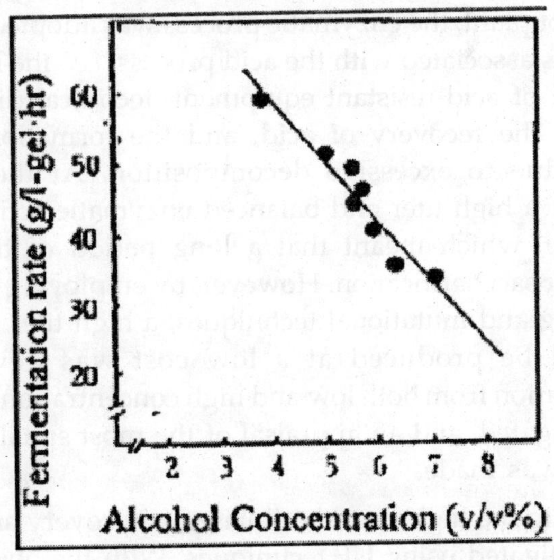

Figure: Effect of alcohol concentration on fermentation rates of immobilised continuous fermentation processes

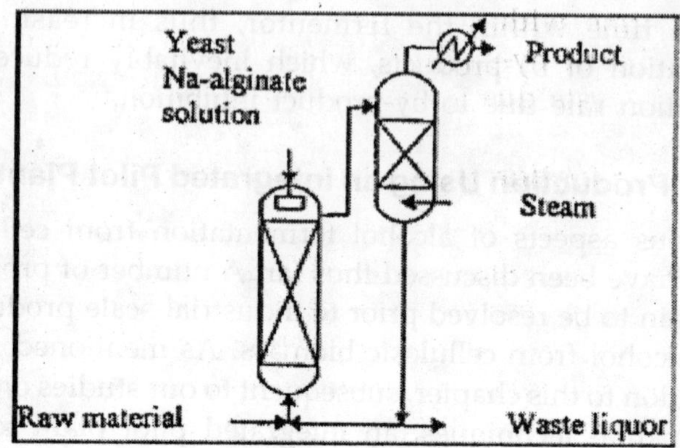

Figure: Schematic of an immobilised flash system

Outline

Bagasse, rice straw, and beech wood chips were used as cellulosic biomass in the pilot plant. These materials vary in composition, and are thus presumed to differ in three-dimensional structure. Cellulose can either be saccharified by an acid process, or by an enzymatic process using cellulase. For the pilot plant, the enzymatic process was adopted in view of problems associated with the acid process, *i.e.*, the necessity for the use of acid-resistant equipment, technical difficulties relating to the recovery of acid, and the formation of by-products due to excessive decomposition. At the outset, cellulose of a high titer and balanced enzymatic activity was unavailable, which meant that a long period of time was required for saccharification. However, by employing a variety of screening and mutational techniques, a high titer cellulose that could be produced at a low cost was developed. Saccharification from both low and high concentration biomass was investigated, and an appraisal of the most suitable types of reactor was made.

Owing to the high cost of cellulase, its recovery and reuse were investigated using UF techniques. With the objective of improving the efficiency of alcohol fermentation and thus saving energy, the use of RO as a means of concentrating the sugar solutions was investigated. Alcohol fermentation was conducted by the immobilised flash method using a bioreactor in combination with the flash method for the purpose of accelerating alcohol production while minimising the influence of alcohol generated on yeast cells. Two methods were used to evaluate the recovery of alcohol:

> The conventional azeotropic distillation method, and a combination of the super critical fluid extraction (SCFE) and per-vaporation methods. In addition, the plant was evaluated as a total system using waste water, which mainly consisted of lignin waste water derived from the pre-treatment step and alcohol fermentation waste water. This waste water was treated by the methane fermentation method using a cell concentration apparatus fitted with a membrane. Low-concentration alcohol leaking from the SCFE-per-vaporation, and immobilised flash fermentation steps was recovered using RO. The processing steps employed in the plant are described individually in the following sections.

Cellulose Production

Bacteria represented by the genus *Cellulomonas,* and yeast strains including those of the genus *Trichosporon* were used for cellulose production. In general, the production of cellulose by fungi was preferred. Various microorganisms obtained domestically as well as from abroad, were screened, and *Trichoderma reesei* QM-9414 isolated by the US Army Natick Laboratory was selected as the best strain. Mutants with enhanced enzymatic titer or with specific desirable characteristics were obtained by mutation using QM-9414 as a parent strain.

The semi-batch culture method produced a high-titer cellulose. A culture method incorporating pH shifting was employed to enhance p-glucosidase activity in *Trichoderma* microorganisms. In addition to scale-up experiments, cheese whey and biomass (bagasse, rice straw, etc.) were investigated as potential inexpensive carbon sources, while soybean meal and residual fermentation cells, were investigated as potential inexpensive nitrogen sources. On the basis of results obtained, a scale-up experiment was carried out in a 1-kL tank using the semi-batch culture method with 10 per cent Avicel (crystalline cellulose) as a carbon source and soybean meal as a nitrogen source, with pH control by the shifting method. High-titer cellulose activities, *i.e.*, 720 U/ml as CMCase, 33 U/ml as FPU, 47 U/ml as Avicelase, and 31 U/ml as p-glucosidase, were obtained. Soluble protein resulting exceeded 5 per cent.

Saccharification of Biomass

High-concentration saccharification was conducted by the batch method. Low-concentration saccharification was performed by the continuous method, the volume of biomass being adjusted to between 3 and 6 per cent. A saccharification rate of 90 to 95 per cent, was obtained after 5 to 7 hours of reaction, with only a small amount of unreacted biomass residue remaining after the reaction. Beech chips treated by steam explosion (Temp: 220°C_2 pressure retention time: 5 min) were also subjected to the saccharification reaction for 24 hours, resulting in a saccharification rate of approximately 75 per cent.

Enzyme Recovery from Biomass

Cellulose encompasses a group of enzymes having different molecular weights. A UF membrane capable of fractionating molecular weights ranging between 10,000 and 20,000 Dalton, was therefore used to recover cellulose for reuse. While enzyme recovery from low-concentration saccharification solutions was approximately 90 per cent, that from high-concentration saccharification solutions was approximately 75 to 80 per cent owing to cellulose adsorption by saccharification residues. Enzyme recovery in the latter case could be markedly improved by reslurring the residue to extract the enzymes. In the pilot plant, tubular-type ultrafiltration equipment with a polystyrene membrane was used.

Concentration of Sugar Solutions

In order to improve the efficiency of alcohol fermentation, UF-filtered sugar solutions which were also free of enzyme were passed through a polyamide, spiral-type RO membrane, enabling them to be concentrated from between 3 and 15 per cent to a final concentration of approximately 30 per cent. Using this loose RO system, more than 95 per cent of the sugar was recovered over an extended period.

Alcohol Fermentation

The immobilised flash method was used for alcohol fermentation in the integrated pilot-scale plant. Using this method, it was possible to maintain the alcohol concentration at approximately 5 per cent, while elevating the sugar concentration of the raw material to as high as 30 per cent. In order to efficiently convert xylose contained in the saccharified solutions of biomass, yeast strains were screened and *Kluyveromyces cellobiovorus*, a strain capable of assimilating cellobiose and producing ethanol by fermentation was newly isolated. This strain was however of poor alcohol resistance, as are other known xylose-assimilating yeast strains. The simultaneous use of the flash method was effective in minimising the influence of alcohol and efficiently inducing the activity of this strain.

Alcohol Recovery

Methods used for recovering alcohol included the distillation method, and the SCFE and osmotic vaporisation methods. In the distillation method, dehydrated ethanol was obtained by ternary azeotropic distillation, with benzene as a solvent. A pressure-type concentrator was employed, with steam at the top of the concentrator as an energy source for the ethanol separation column. Using this methodology, energy savings of 30 per cent over that of the conventional method were achieved.

In the SCFE, ethanol obtained from the flash distillation step, ranging in concentration from 20 to 30 per cent was brought into çontact with supercritical carbon dioxide gas, resulting in 90 per cent recovery of 85 to 95 per cent concentrated ethanol solutions. By optimising the extraction pressure and temperature and, at the same time, providing a mist separator at the top of the extractor, ethanol having a composition approaching that of the azeotrope was recoverable. Impurities in the culture broth exhibited the same behaviour as that of the ethanol. Studies on osmotic vaporisation membrane methods for dehydration, resulted in alcohol concentrations exceeding 99.5 per cent at a recovery rate of at least 95 per cent. The water content of alcohol recovered, was less than 0.5 per cent, indicating that this alcohol may be feasibly used as fuel ethanol. Dehydrated ethanol produced from bagasse was mixed with gasoline and its performance as a fuel for automobiles was evaluated by Cosmo Oil Co., Ltd. Results showed no substantial difference from ordinary reagent ethanol, and the ethanol was deemed suitable for practical use.

Important Areas in Photosynthesis

Bisphosphoglycerate

1,3-Bisphosphoglycerate (1,3BPG), also known as PGAP, is a 3-carbon organic molecule present in most, if not all, living organisms. It primarily exists as a metabolic intermediate in both glycolysis during respiration and the Calvin cycle during photosynthesis. 1,3BPG is a transitional stage between glycerate 3-phosphate and glyceraldehyde 3-phosphate during the fixation/reduction of CO_2. 1,3BPG is also a precursor to 2,3-bisphosphoglycerate which in turn is a reaction intermediate in the glycolytic pathway.

Biological Structure and Role

1,3-bisphosphoglycerate is the anion form of bisphosphoglyceric acid. It is phosphorylated at the number 1 and 3 carbons. The result of this phosphorylation gives 1,3BPG important biological properties such as the ability to phosphorylate ADP to form the energy storage molecule ATP.

In Glycolysis

1,3BPG is a metabolic intermediate in the glycolytic pathway. It is created by the exergonic oxidation of the aldehyde in G3P. The result of this oxidation is the conversion of the aldehyde group into a carboxylic acid group which drives the formation of an acyl phosphate bond. This is incidentally the only step in the glycolytic pathway in which NAD^+ is converted into NADH. The formation reaction of 1,3BPG requires the presence of an enzyme called glyceraldehyde-3-phosphate dehydrogenase.

The high energy acyl phosphate bond of 1,3BPG is important in respiration as it assists in the formation of ATP. The molecule of ATP created during the following reaction is the first molecule produced during respiration. The reaction occurs as follows;

1,3-bisphosphoglycerate + ADP ⟶ 3-phosphoglycerate + ATP

The transfer of an inorganic phosphate from the carboxyl group on 1,3BPG to ADP to form ATP is reversible due to a low ΔG. This is as a result of one acyl phosphate bond being cleaved whilst another is created. This reaction is not naturally spontaneous and requires the presence of a catalyst. This role is performed by the enzyme phosphoglycerate kinase. During the reaction phosphoglycerate kinase undergoes a substrate induced conformational change similar to another metabolic enzyme called hexokinase.

Because two molecules of glyceraldhye-3-phosphate are formed during glycolysis from one molecule of glucose, 1,3BPG can be said to responsible for two of the ten molecules of ATP produced during the entire process. Glycolysis also uses two molecules of ATP in its initial stages as a committed and irreversible step. For this reason glycolysis is not reversible and has a net produce of 2 molecules of ATP and two of NADH. The two molecules of NADH themselves go on to produce approximately 3 molecules of ATP each.

In the Calvin Cycle

1,3-BPG has a very similar role in the Calvin cycle to its role in the glycolytic pathway, for this reason both reactions are said to be analogous. However the reaction pathway is effectively reversed.

The only other major difference between the two reactions is that NADPH is used as an electron donor in the calvin cycle whilst NAD^+ is used as an electron acceptor in glycolysis. In this reaction cycle 1,3BPG originates from 3-phosphoglycerate and is made into glyceraldehyde 3-phosphate by the action of specific enzymes.

Contrary to the similar reactions of the glycolytic pathway, 1,3BPG in the calvin cycle does not produce ATP but instead uses it. For this reason it can be considered to be an irreversible and committed step in the cycle. The outcome of this section of the cycle is an inorganic phosphate is removed from 1,3BPG and used to provide energy for the donation of electrons from NADPH into the reaction cycle, forming $NADP^+$.

In complete reverse of the glycolytic pathway reaction, the enzyme phosphoglycerate kinase catalyses the reduction of the carboxyl group of 1,3BPG to form an aldehyde instead.

This reaction also releases an inorganic phosphate molecule which is subsequently used as energy for the donation of electrons from the conversion of NADPH to $NADP^+$. Overseeing this latter stage of the reaction is the enzyme glyceraldehyde-phosphate dehydrogenase.

Medical significance of 1,3BPG

During normal metabolism in humans approximately 20 per cent of the 1,3BPG produced does not go any further in the glycolytic pathway. It is instead shunted through an alternate pathway involving the reduction of ATP in the erythrocytes. During this alternate pathway it is made into a similar molecule called 2,3-Bisphoshoglycerate. 2,3BPG is used as a mechanism to oversee the efficient release of oxygen from haemoglobin.

Levels of this 1,3BPG will raise in a patients blood when oxygen levels are low as this is one of the mechanisms of acclimatisation. Low oxygen levels trigger a rise in 1,3BPG levels which in turn raises the level of 2,3BPG which alters the efficiency of oxygen dissociation from haemoglobin. This constitutes a negative feedback relationship.

Mechanism of 1,3-bisphosphoglycerate transfer from phosphoglycerate kinase to glyceraldehyde-3-phosphate dehydrogenase

The kinetics of 1,3-bisphosphoglycerate binding to glyceraldehyde-3- phosphate dehydrogenase have been examined by stopped-flow techniques in the absence and presence of phosphoglycerate kinase, using enzyme concentrations in the range 0.5-40 µm. Rate and equilibrium constant estimates for the interaction of the ligand with the two enzymes are reported. The kinetics of ligand transfer from the binary complex of bisphosphoglycerate and phosphoglycerate kinase to the binary complex of NAD+ and glyceraldehyde-3-phosphate dehydrogenase conform excellently to the predictions of a standard free-diffusion mechanism and exhibit no detectable contributions from a mechanism of direct (channelised) transfer of bisphosphoglycerate between the two enzymes. Previously reported evidence that the binary complex of bisphosphoglycerate and phosphoglycerate kinase may act (in the presence of NADH) as a substrate for glyceraldehyde-3-phosphate dehydrogenase according to Michaelis-Menten kinetics is based on a misinterpretation of the experimental observations that can be attributed to neglect of the autocatalytic effect of NAD+ produced during the reaction. Experiments performed under conditions where the autocatalytic effect of NAD+ is eliminated provide clear evidence that the kinetics of utilisation of the kinase-bisphosphoglycerate complex for enzymic NADH reduction are consistent with prior dissociation of the complex according to a free-diffusion mechanism of metabolite transfer and incompatible with a mechanism of direct metabolite transfer. A kinetic argument is presented which renders implausible the very idea that direct metabolite transfer between 'soluble' consecutive enzymes in metabolic pathways may offer any catalytic advantages in comparison to metabolite transfer by free diffusion.

A mechanism of direct metabolite transfer seems intuitively attractive only because one tends to disregard the diffusional processes required to bring the consecutive enzymes together and to separate them when the transfer has been completed. Direct metabolite transfer would be expected to be catalytically advantageous only in tightly bound multienzyme complexes showing no kinetically significant tendency to dissociate. It is concluded that mechanisms of direct metabolite transfer have not been convincingly demonstrated to apply, nor are they likely to apply, between 'soluble' consecutive enzymes in metabolic pathways, at least not in the glycolytic sequence of reactions.

Bundle-sheath

Bundle-sheath cells are contained in certain C_4 plants. These plants use C_4 carbon fixation to carry out photosynthesis when normal conditions for C_3 plants aren't present (e.g. normal water levels, CO_2 levels, heat, light).

Bundle-sheath cells are photosynthetic cells arranged into tightly packed sheaths around the veins of a leaf.

They form a protective covering these leaf veins, and consist of one or more cell layers, usually parenchyma. Between the bundle-sheath and the leaf surface are the more loosely arranged mesophyll cells. The Calvin cycle is confined to the chloroplasts of these bundle-sheath cells.

Compensation Point

The compensation point is the point during photosynthesis where the rate of photosynthesis exactly matches the rate of respiration; *i.e.* the input of CO_2 or O_2 is the same as the output of the other by a plant. This point is reached during early mornings and late evenings. It is also involved in sugar production.

This is a summary of compensation point.

- The rate of photosynthesis is exactly equal to the rate of respiration.
- The amount of oxygen produced by a plant is equal to the amount used up at that point of time.
- Thus there is no net output of oxygen by the plant.

Crassulacean Acid Metabolism

Crassulacean acid metabolism, also known as CAM photosynthesis, is an elaborate carbon fixation pathway in some photosynthetic plants.

CAM is usually found in plants living in arid conditions, including those found in the desert (for example, cacti or pineapple). It is named after the plant family it was first discovered in, the Crassulaceae (the jade plants, sedums, and relatives).

Synopsis

Plants that have adaptations allowing them to thrive in dry climates are called xerophytes. Some xerophytic plants have small, thick leaves with a reduced surface area to volume ratio. They may also have a thickened cuticle to protect themselves from drying out.

The stomata may be sunken into pits. Some xerophytes shed their leaves during the dry season and others, such as cacti, orchids, and bromeliads can store water in vacuoles. Some xerophytes perform photosynthesis using Crassulacean acid metabolism (or CAM).

CAM plants close their stomata during the day in order to conserve water by preventing evapotranspiration. Their stomata then open during the cooler and more humid night-time hours, allowing uptake of carbon dioxide for use in carbon fixation.

This is begun when the three-carbon compound phosphoenolpyruvate is carboxylated into oxaloacetate which is then reduced to form malate. CAM plants store these four-carbon intermediates and other simple organic compounds in their vacuoles.

Malate is easily broken down into pyruvate and CO_2, after which pyruvate is phosphorylated to regenerate phosphoenolpyruvate (PEP). In the daytime, the malic acid is removed from the vacuoles and cleaved to produce CO_2 so that it can be utilised by the enzyme Rubisco in the Calvin-Benson cycle in the chloroplast stroma. Thus, by reducing evapotranspiration rates during gas exchange, CAM allows plants to grow in environments that would otherwise be far too dry for plant growth or, at best, subject them to severe drought stress.

In some ways, CAM resembles C_4 photosynthesis, except that CAM plants contain no bundle-sheath cells. C_4 plants capture the CO_2 in one type of cell tissue (mesophyll) and then transfer it to another type of tissue (bundle-sheath cells) so that carbon fixation may occur via the Calvin-Benson cycle.

Furthermore, C_4 metabolism is continuous as long as there is light available, while CAM occurs only at night. Thus, C_4 metabolism physically separates CO_2 fixation from the Calvin cycle, while CAM metabolism temporally separates CO_2 fixation from the Calvin cycle.

CAM plants are very good at retaining water, and are also very efficient with the usage of nitrogen. They are inefficient at absorbing CO_2, however, they are slow growing compared to other plants. In addition, CAM plants avoid photorespiration.

The enzyme responsible for fixing carbon into the Calvin cycle, Rubisco, cannot distinguish CO_2 from oxygen. As a result, the plant uses energy to break down carbon compounds. This costly process occurs when the internal concentration of oxygen in the leaves is too high, particularly in C_3 plants.

C_4 Carbon Fixation

C_4 carbon fixation is one of three methods, along with C3 and CAM photosynthesis, used by land plants to "fix" carbon dioxide (binding the gaseous molecules to dissolved compounds inside the plant) for sugar production through photosynthesis. Along with CAM photosynthesis, C_4 fixation is an improvement over the simpler and more ancient C_3 carbon fixation strategy used by most plants. Both methods overcome the tendency of Rubisco (the first enzyme in the Calvin cycle) to photorespire, or waste energy by using oxygen to break down carbon compounds to CO_2. C_4 plants separate Rubisco from atmospheric oxygen, fixing carbon in the mesophyll cells and using oxaloacetate and malate to ferry the fixed carbon to Rubisco and the rest of the Calvin cycle enzymes isolated in the bundle-sheath

The Pathway

The C_4 pathway was discovered by M. D. Hatch and C. R. Slack, two Australian researchers, in 1966, so it is sometimes called the Hatch-Slack pathway.

In C_3 plants, the first step in the light-independent reactions of photosynthesis involves the fixation of CO_2 by Rubisco enzyme into 3-phosphoglycerate. However, due to the dual carboxylase/oxygenase activity of Rubisco, an amount of the substrate is oxidised rather than carboxylated resulting in loss of substrate and consumption of energy, in what is known as photorespiration. In order to bypass the photorespiration pathway, C_4 plants have developed a mechanism to efficiently deliver CO_2 to the Rubisco enzyme. They utilise their specific leaf anatomy where chloroplasts exist not only in the mesophyll cells in the outer part of their leaves but in the bundle-sheath cells as well. Instead of direct fixation in the calvin cycle, CO_2 is converted to a 4-carbon organic acid which has the ability to regenerate CO_2 in the chloroplasts of the bundle sheath cells. Bundle sheath cells can then utilise this CO_2 to generate carbohydrates by the conventional C_3 pathway.

The first step in the pathway is the fixation of CO_2 by the enzyme phosphoenolpyruvate carboxylase which exists in the mesophyll cells:

PEP carboxylase + PEP + $CO_2 \longrightarrow$ oxaloacetate

PEP carboxylase has a lower Km for CO_2 and hence higher affinity- than Rubisco. Furthermore, O_2 is a very poor substrate for this enzyme. Thus, at relatively low concentrations of CO_2, most CO_2 will be fixed by this pathway.

The product is usually converted to malate, a simple organic compound that is transported to the bundle-sheath cells surrounding a nearby vein, where it is decarboxylated to release CO_2, which enters Calvin cycle. The decarboxylation leaves pyruvate, which is transported back to the mesophyll and phosphorylated, in a reaction catalysed by pyruvate, orthophosphate dikinase (PPDK), to regenerate PEP at the cost of a phosphorus group and one ATP molecule.

Since every CO_2 molecule has to be fixed twice, the C_4 pathway is more energy-consuming than the C_3 pathway. The C_3 pathway requires 18 ATP for the synthesis of one molecule of glucose while the C_4 pathway requires 30 ATP. But since otherwise tropical plants lose more than half of photosynthetic carbon in photorespiration, the C_4 pathway is an adaptive mechanism for minimising the loss.

There are several variants of this pathway:

1. The 4-carbon acid transported from mesophyll cells may be malate as above, or may be aspartate.
2. The 3-carbon acid transported back from bundle-sheath cells may be pyruvate as above, or alanine.
3. The enzyme which catalyses decarboxylation in bundle-sheath cells differs. In maize and sugarcane, the enzyme is NADP-malic enzyme, in millet, it is NAD-malic enzyme, and in *Panicum maximum* it is PEP carboxykinase.

C_4 Leaf Anatomy

The C_4 plants possess a characteristic leaf anatomy. Their vascular bundles are surrounded by two rings of cells. The inner ring, called Bundle Sheath Cells, contain starch-rich chloroplasts lacking grana which differ from those in mesophyll cells present as the outer ring. Hence, the chloroplasts are called dimorphic. This peculiar anatomy is called Kranz Anatomy (Kranz-Crown/Halo). The primary function of the Kranz is to provide a site in which carbon dioxide can be concentrated around Rubisco, thus reducing photorespiration. In order to facilitate the maintenance of a significantly higher carbon dioxide concentration in the bundle sheath compared to the mesophyll, the boundary layer of the Kranz has a low conductance to carbon dioxide, a property which may be enhanced by the presence of suberin.

Although most C_4 plants exhibit Kranz anatomy, there are a number of species which operate a limited C_4 cycle without any distinct bundle sheath tissue. *Suaeda aralocaspica* (formerly known as *Borszczowia aralocaspica*), *Bienertia cycloptera and Bienertia sinuspersici* are terrestrial plants which inhabit dry, salty depressions in the deserts of South East Asia. These plants have been shown to operate single-cell C_4 carbon dioxide concentrating mechanisms which are unique amongst the known C_4 mechanisms. Although the cytology of both species differ slightly, the basic principle is that fluid filled vacuoles are employed to separate the cell into separate areas. Carboxylation enzymes in the cytosol can therefore be kept separate from decarboxylase enzymes and Rubisco in the chloroplasts, and a diffusive barrier can be established between the chloroplasts (which contain Rubisco) and the cytosol. This enables a bundle-sheath type area and a mesophyll type area to be established within a single cell. Although this does allow a limited C_4 cycle to operate, it is relatively inefficient, with much leakage of CO_2 from around Rubisco occurring. There is also evidence for the non-Kranz aquatic macrophyte Hydrilla verticillata exhibiting inducible C_4 photosynthesis under warm conditions, although the mechanism by which CO_2 leakage from around Rubisco is minimised is currently uncertain.

Evolution and Advantages of the C_4 Pathway

C_4 plants have a competitive advantage over plants possessing the more common C_3 carbon fixation pathway under conditions of drought, high temperatures and nitrogen or carbon dioxide limitation. Carbon fixation has evolved on at least 18 independent occasions in different groups of plants, so is an example of convergent evolution. Plants which use C_4 metabolism include sugarcane, maize, sorghum, *Eleusine, Amaranthus,* and Switchgrass (*Panicum virgatum*). C_4 plants arose during the Cenozoic Era and did not become common until the Miocene Period. Today they represent about 5 per cent of earth's plant biomass and 1 per cent of its known plant species. These species are concentrated in the tropics where the high air temperature contributes to higher possible levels of oxygenase activity by Rubisco, which increases rates of photorespiration in C_3 plants.

C_3 Carbon Fixation

C_3 carbon fixation is a metabolic pathway for carbon fixation in photosynthesis. This process converts carbon dioxide and ribulose bisphosphate (RuBP, a 5-carbon sugar) into 3-phosphoglycerate through the following reaction:

$$6\,CO_2 + 6\,\text{RuBP} \longrightarrow 12\ \text{3-phosphoglycerate}$$

This reaction occurs in all plants as the first step of the Calvin cycle. In C_4 plants, carbon dioxide is drawn out of malate and into this reaction rather than directly from the air.

Plants that survive solely on C_3 fixation (C_3 plants) tend to thrive in areas where sunlight intensity is moderate, temperatures are moderate, carbon dioxide concentrations are around 200 ppm or higher, and ground water is plentiful. The C_3 plants, originating during Mesozoic and Paleozoic era, predate the C_4 plants and still represent approximately 95 per cent of earth's plant biomass.

C_3 plants must be in areas with high concentrations of carbon dioxide because Rubisco often incorporates an oxygen molecule into the RuBP, instead of a carbon dioxide molecule. This breaks the RuBP into a three-carbon sugar that can remain in the Calvin cycle, and two molecules of glycolate which is oxidised into carbon dioxide, wasting the cell's energy. High concentrations of carbon dioxide lowers the chance that Rubisco incorporates an oxygen molecule. C_4 and CAM plants have adaptations that allow them to survive in areas where the plant cannot take in a lot of carbon dioxide.

The isotopic signature of C_3 plants shows higher degree of ^{13}C depletion than the C_4 plants.

Cytochrome b_6f Complex

The cytochrome b_6f complex (plastoquinol—plastocyanin reductase; EC 1.10.99.1) of chloroplasts and cyanobacteria transfers electrons between the two reaction centre complexes of oxygenic photosynthetic membranes, photosystem I and photosystem II, and participates in formation of the transmembrane electrochemical proton gradient by also transferring protons from the stromal to the internal lumen compartment. It is minimally composed of four subunits: cytochrome b_6, carrying a low- and a high-potential haem groups (b_L and b_H); cytochrome *f* with one covalently bound haem *c*; Rieske iron-sulphur protein (ISP) containing a single [Fe_2S_2] cluster; and subunit IV (17 kDa protein). In its structure and functions, the cytochrome b_6f complex bears extensive analogy to the cytochrome bc_1 complex of mitochondria and photosynthetic purple bacteria. However, there are important differences between the two complexes:

- The single-polypeptide cytochrome *b* in the cytochrome bc_1 complex corresponds to cytochrome b_6 and subunit IV in the cytochrome b_6f complex.
- Cytochrome *f* and cytochrome c_1 are not homologous
- The cytochrome b_6f complex contains additional

chromophores, chlorophyll *a*, a-carotene and atypical haem c_i (haem *x*), the latter being linked by a single thioether bond to cytochrome b_6.

The cytochrome b_6f complex is responsible for "non-cyclic" (1) and "cyclic" (2) electron transfer between two mobile redox carriers, plastoquinol (QH_2) and plastocyanin:

H_2O → photosystem II → QH_2 → Cyt b_6 f → plastocyanin → photosystem I → $NADP^+$ (1)

QH_2 → Cyt b_6f → plastocyanin → photosystem I → Q (2)

Electron transfer is coupled with the translocation of protons across the membrane, thus generating proton-motive force in the form of an electrochemical proton potential which can drive ATP synthesis. The crystal structure of cytochrome b_6f complexes from *Chlamydomonas reinhardtii* and *Mastigocladus laminosus* have been determined.

Cytochrome *f*

Cytochrome *f* is the largest subunit of cytochrome b_6f complex (plastoquinol—plastocyanin reductase; EC 1.10.99.1). The crystal structures of soluble (lumen-side) domain of cytochrome *f* reveal two structural domains: a small one above a larger one which, in turn, is on top of the attachment to the membrane domain. The large domain consists of an anti-parallel β-sandwich and a short haem-binding peptide, which form a three-layer structure. The small domain is inserted between two β-strands of the large domain and is an all-β domain.

The haem is bound between two short helices at the N-terminus of cytochrome *f*. Within the second helix is the sequence motif for the *c*-type cytochromes, Cys-Xaa-Xaa-Cys-His, which is covalently attached to the haem through thioether bonds of the two cysteine residues. The fifth haem iron ligand is histidine and the sixth haem iron ligand is the á-amino group of N-terminal tyrosine residue (Tyr-1).

Cytochrome *f* has an internal network of water molecules. It has been hypothesised that this water chain functions as a proton wire. The water chain appears to be a conserved feature of cytochromes *f*.

Dole Effect

The Dole effect describes an inequality in the ratio of the heavy isotope ^{18}O (a 'standard' oxygen atom with two additional neutrons) to the lighter ^{16}O, measured in the atmosphere and seawater. It was noticed in 1935 that air contained relatively more ^{18}O than seawater; this was quantified in 1975 to 23.5 per cent. The imbalance arises mainly as a result of respiration in plants and animals. Due to thermodynamics, respiration removes the lighter - hence more reactive - ^{16}O in preference to ^{18}O, increasing the relative amount of ^{18}O in the atmosphere.

The inequality is balanced by photosynthesis. Photosynthesis emits oxygen with the same isotopic composition (i.e. the ratio between ^{18}O and ^{16}O) as the water (H_2O) used in the reaction, which is independent of the atmospheric ratio. Thus when atmospheric ^{18}O levels are high enough, photosynthesis will act as a reducing factor. However, as a complicating factor, the degree of fractionation (i.e. change in isotope ratio) occurring due to photosynthesis is not entirely dependent on the water drawn up by the plant, as fractionation can occur as a result of preferential evaporation of $H_2{}^{16}O$-water bearing lighter oxygen isotopes, and other small but significant processes.

Use of Dole Effect

Since evaporation causes oceanic and terrestrial waters to have a different ratio of ^{18}O to ^{16}O, the Dole Effect will reflect the relevant importances of land-based and marine photosynthesis. The complete removal of land-based productivity would result in a Dole effect shift of 2.3 per cent from the current value of 23.5 per cent.

The stability (to within 0.5 per cent) of the atmospheric ^{18}O

to ^{16}O ratio with respect to sea surface waters during the Palaeocene (the last 130,000 years), as derived from ice cores, suggests that terrestrial and marine productivity have varied together during this time period.

The Dole effect can also be applied as a tracer in sea water, with slight variations in chemistry being used to track a discrete 'parcel' of water and determine its age.

Primary Production and Vertical Export

Primary production is the production of organic compounds from atmospheric or aquatic carbon dioxide, principally through the process of photosynthesis, with chemosynthesis being much less important. All life on earth is directly or indirectly reliant on primary production. The organisms responsible for primary production are known as primary producers or autotrophs, and form the base of the food chain. In terrestrial ecosystems, these organisms are mainly plants, while in aquatic ecosystems algae are primarily responsible. Primary production is distinguished as either net or gross, the former accounting for losses to processes such as cellular respiration, the latter not.

At the most fundamental level, primary production is the capture of energy in the form of electromagnetic radiation and its conversion to, and storage as, chemical energy by living organisms. The main source of this energy is the sun. A very tiny fraction of primary production is driven by organisms utilising the chemical energy of inorganic molecules.

Regardless of its source, this energy is used to synthesise complex organic molecules from simpler inorganic compounds such as carbon dioxide (CO_2) and water (H_2O). The following two equations are simplified representations of photosynthesis (top) and (one form of) chemosynthesis (bottom) :

$$CO_2 + H_2O + \textit{light} \longrightarrow CH_2O + O_2$$

$$CO_2 + O_2 + 4\,H_2S \longrightarrow CH_2O + 4\,S + 3\,H_2O$$

In both cases, the end point is reduced carbohydrate (CH_2O), typically molecules such as glucose or other sugars. These relatively simple molecules may be then used to synthesise further more complicated molecules, including proteins, complex carbohydrates, lipids, and nucleic acids, or be respired to perform work. Consumption of primary producers by heterotrophic organisms, such as animals, then transfers these organic molecules (and the energy stored within them) up the food chain, fuelling all of the earth's living systems.

Terrestrial Production

On the land, almost all primary production is now performed by vascular plants, although a small fraction comes from algae and non-vascular plants such as mosses and liverworts. However, before the evolution of vascular plants, non-vascular plants played a more significant role. Primary production on land is a function of many factors, but principally local hydrology and temperature (the latter covaries to an extent with light, the source of energy for photosynthesis). While plants cover much of the earth's surface, they are strongly curtailed wherever temperatures are too extreme or where necessary plant resources (principally water and light) are limiting, such as deserts or polar regions.

Water is "consumed" in plants by the processes of photosynthesis and transpiration. The latter process (which is responsible for about 90 per cent of water use) is driven by the evaporation of water from the leaves of plants. It allows plants to transport water and mineral nutrients from the soil to growth regions, and also cools a plant down. It can be regulated by structures known as stomata, but these also regulate the supply of carbon dioxide from the atmosphere, so that decreasing water loss also decreases carbon dioxide gain. Crassulacean acid metabolism (CAM) and C_4 plants use physiological and anatomical workarounds to increase their water-use efficiency and allow increased primary production to take place under conditions that would limit "normal" C_3 plants (the majority).

Primary production consists of *new production* (P_N) that is based on allochthonous, i.e. externally supplied nutrients, and *regenerated production* (P_R), which is based on autochthonous, i.e. internally recycled nutrients. Hence total primary production (P_T) is the sum of P_N and P_R. The amount of carbon that enters the aphotic zone is entitled *export production* (P_E).

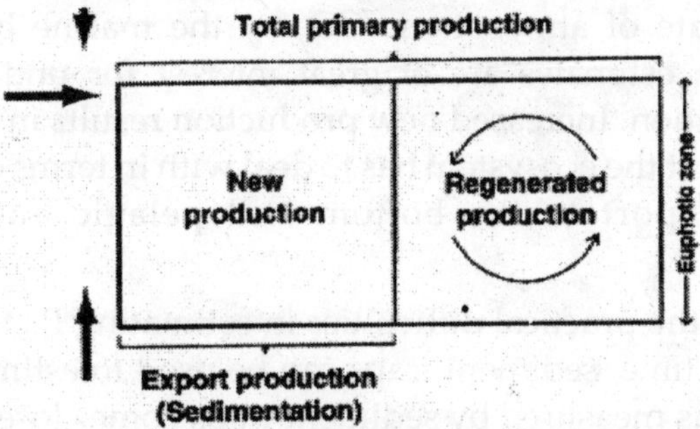

Figure: New and regenerated production are based on (a) the supply of the limiting (allochthonous) nutrients from the aphotic zone, by advection, run-off or from the atmosphere (straight arrows) and (b) the recycled (autochthonous) nutrients in the euphotic zone (circular arrows), respectively. New and regenerated production comprise total primary production. Export production is the amount of sinking organic carbon at the bottom of the euphotic zone.

The concept of new production is of utmost importance for understanding natural and eutrophicated ecosystems because the fraction $f = P_N/P_T$ represents the upper limit of organic matter and energy which can be removed or extracted from the surface waters of the system without destroying the long-term integrity of pelagic systems. P_N represents thus the biomass that has to be handled by an eutrophicated recipient (e.g. mineralisation, accumulation, harvestable biomass and export of biomass to adjacent recipients). Given the importance of P_N for the overall cycling of organic matter, considerable emphasis has recently been given to estimating P_N in coastal (e.g. Wassmann) as well as oceanic environments (e.g. Knauer).

New production represents the carrying capacity of a marine ecosystem. New production represents the maximum production capacity of an ecosystem or the harvestable production. New production is a critical component of marine primary production that limits the supply of food to the benthos, zooplankton, fish, and extensive aquaculture as well as the removal rate of atmospheric CO_2 by the marine biota. New production estimates are of great interest for understanding eutrophication. Increased new production results in additional biomass that the ecosystem has to deal with in terms of grazing, vertical export to the bottom and pelagic and benthic degradation.

Given the practical difficulties in estimating P_N for lengthy periods of time, sediment traps can be used to estimate P_N. P_E estimates as measured by sediment traps come close to P_N, but are always smaller because it comprises only the particulate fraction and some transformation takes place from ammonium to nitrate even in the upper layers. Calculations of the productivity index f by applying P_E give, therefore, rise to underestimates.

As a consequence, the term $e = P_E/P_T$ can be applied and used as an approximation of f. In boreal, coastal areas where steady state, if at all, cannot be assumed for intervals of less than the length of the productive period, e is meaningful as a base for estimating f for lengthy periods only (e.g. >6 months). Therefore, the term <e>, representing e for lengthy periods of time will be applied.

Nutrient Supply, Primary Production, Retention and Vertical Export

Increased supply of nutrients to the euphotic zone gives rise to increased production of algae which sooner or later sink to deeper water and the sediment, resulting in increased sediment-water exchange rates, at times in mass mortality of macrofauna and fish eggs and finally in anoxia.

During the last decades widespread occurrence of low

oxygen concentrations or anoxia in bottom waters, decreased catches of fish and blooms of toxic algae threatening aquaculture as well as stocks of wild animals have been reported with increasing frequency.

These changes seem to be caused by increased inputs of nutrients to aquatic areas from sewage, agricultural run-off and atmospheric fall-out, giving rise to various degrees of eutrophication of freshwater as well as marine, coastal environments (e.g. Wulff).

Figure below shows the principle processes of atmospheric CO_2 uptake and release, primary production, suspended biomass and vertical export to the bottom. Seawater takes up CO_2 from the atmosphere that is either taken up by phytoplankton or released again to the atmosphere. The phytoplankton uptake of CO_2 is caused by primary production and first and foremost dependent on photosynthetic active radiation and nutrients.

In addition it is influenced by the residence time of phytoplankton in the euphotic layer (determined by vertical mixing and stratification). Phytoplankton accumulates in the upper layers if grazing and degradation rates are lower than primary production, i.e. a bloom takes shape. A part of the suspended biomass consistent on phytoplankton cells and detritus from grazing and degradation will sink into the aphotic zone and further to the bottom.

The export of biogenic matter to the bottom is a complex function of the total amount of suspended matter, the sinking velocity of the sinking particles and the degradation impact of the pelagic heterotrophs. For example, low suspended biomass, low pelagic degradation and high sinking rates give result in a similar sedimentation rate at depth than high suspended biomass, high pelagic degradation and low sinking rates. Degradation of organic matter in the water column or the sediment results in nutrients that sooner or later can be taken up by phytoplankton.

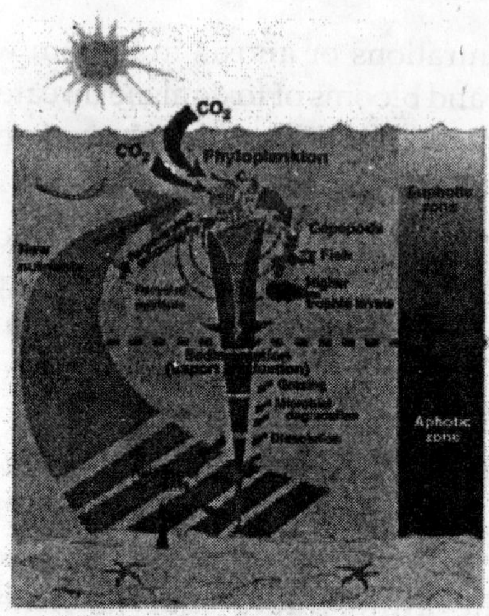

Figure: Primary production, vertical flux and regeneration of nutrients in a coastal marine ecosystem. Also shown are some of the involved organisms such as phytoplankton, zooplankton, higher trophic levels and benthic organisms. The massive and narrow vertical arrows indicate scenarios of substantial and insignificant vertical flux.

If the limiting nutrient is nitrogen the net primary production depends on the allochthonous nutrient nitrate while the remaining primary production is based upon the autochthonous nutrient ammonium that derives from internal recycling by heterotrophic organisms.

The basic principle to use nitrogen species to determine how much of the total primary production comprises new production (in case nitrogen is the limiting element) is difficult to apply in shallow water where the cycle of nutrients is rapid and where particulate nitrogen supplied to the bottom can be recycled to nitrate that is available for primary production. In this case nitrate is not new, but regenerated. Thus some of the nitrate is not 'new' and does not comply with the basic assumption of new production. In countless eutrophicated regions ammonium and urea are supplied as allochthonous nutrients. Also in this case the traditional method to distinguish new from regenerated production is not possible because some of the per definition autochthonous nutrients are allochthonous. New production is thus impossible to measure in eutrophicated waters.

The state of an ecosystem during a transient bloom is basically characterised by export food chains with high vertical export. The amount of regenerated production increases, as the planktonic system develops and becomes more complex during the post-bloom phase. Sedimentation of organic material is low and the ecosystem is characterised by retention food chains. In the non-eutrophicated coastal zone export chains are based upon new production and represent episodic events on the background of a continuous, seasonally variable regenerated production based on the recycled nutrients from retention chains. If eutrophication continues, i.e. nutrients are supplied in a steady manner, a new steady state with a mixture of export and retention food chains will develop.

In conclusion, an estimate of new/net/harvestable production as a consequence of eutrophication in coastal zones is difficult to measure, among other reason because our terminology and measuring techniques are inadequate.

Gross Primary Production

Primary production is the fixation of light energy into chemical compounds (the construction of sugars from CO_2 and water; photosynthesis). Gross primary production then refers to the total amount of energy fixed by plants (the primary producers).

This energy can then be used by plants to generate new biomass; growing. A certain amount of the ecosystems gross primary production can be consumed by herbivores (primary consumers) without destroying the plant community, this balance is however rather delicate.

FAPAR (Fraction of Absorbed Photosynthetically Active Radiation)

FAPAR, sometimes also noted FAPAR or FPAR is the fraction of the incoming solar radiation in the Photosynthetically Active Radiation spectral region that is absorbed by plants. This biophysical variable is directly related to the primary productivity of plants and some models use it to estimate the assimilation of carbon dioxide in vegetation.

FAPAR can be derived from space measurements in the solar spectral range and a number of state of the art algorithms have been proposed to estimate this important environmental variable. FAPAR can also be used as an indicator of the state and evolution of the vegetation cover; in this function it advantageously replaces the Normalised Difference Vegetation Index (NDVI), provided it is itself properly estimated.

FAPAR is one of the Essential Climate Variables recognised by the UN Global Climate Observing System (GCOS) as necessary to characterise the climate of the earth. GCOS has issued specific recommendations to monitor this variable systematically, both through a reanalysis of existing databases and in the future with current and forthcoming instruments.

Photosynthetically Active Radiation

The expression Photosynthetically Active Radiation, often abbreviated PAR, designates the spectral range of solar light from 400 to 700 nanometres that is useful to terrestrial plants in the process of photosynthesis. This spectral region corresponds more or less with the range of light visible to the human eye. Photons at shorter wavelengths tend to be so energetic that they can be damaging to cells and tissues; fortunately they are mostly filtered out by the ozone layer in the stratosphere. Photons at longer wavelengths do not carry enough energy to allow photosynthesis to take place, and plants have developed, through billions of years of evolution, the capacity to scatter these photons away, hence the very high reflectance and transmittance of live green leaves.

Other living organisms, such as green bacteria, purple bacteria and Heliobacteria, can exploit solar light in slightly extended spectral regions, such as the near-infrared. These bacteria live in environments such as the bottom of stagnant ponds, sediment and ocean depths. Because of their pigments, they form colourful mats of green, red and purple. These organisms must make use of the leftovers discarded by the plant kingdom, in this case, light outside the PAR range.

Chlorophyll, the most abundant plant pigment, is most efficient in capturing red and blue light. Horticulturists say that blue light is the most important for leaf growth and that red light encourages flowering. Accessory pigments such as carotenes and xanthophylls harvest some green light and pass it on to the photosynthetic process, but enough of the green is reflected to give leaves their characteristic colour. An exception to the predominance of chlorophyll is autumn, when chlorophyll decays earlier than the accessory pigments that remain to colour the leaves red, yellow and orange.

PAR measurement is used in agriculture, forestry and oceanography. One of the requirements for productive farmland is adequate PAR, so PAR is used to evaluate agricultural investment potential. PAR sensors stationed at various levels of the forest canopy measure the pattern of PAR availability

Ferredoxin

Ferredoxins (from Latin *ferrum*: iron + redox, often abbreviated "fd") are iron-sulphur proteins that mediate electron transfer in a range of metabolic reactions.

Overview

The term "ferredoxin" was coined by D.C. Wharton of the DuPont Co. and applied to the "iron protein" first purified in 1962 by Mortenson, Valentine and Carnahan from the anaerobic bacterium *Clostridium pasteurianum* (Valentine, 1964). Another redox protein, isolated from spinach chloroplasts by Tagawa and Arnon in 1962, was termed "chloroplast ferredoxin". The chloroplast ferredoxin is involved in both cyclic and non-cyclic photophosphorylation reactions of photosynthesis. In non-cyclic photophosphorylation, ferredoxin is the last electron acceptor, and reduces the enzyme NADP+ reductase. It "picks up" electrons produced from sunlight-excited chlorophyll and transfers them to the enzyme ferredoxin:$NADP^+$ oxidoreductase EC 1.18.1.2. Ferredoxins are small proteins containing iron and sulphur atoms organised as iron-sulphur clusters. These biological "capacitors" can accept or discharge electrons, the effect being change in the oxidation states (+2 or +3) of the iron atoms. This way, ferredoxin acts as electron transfer agents in biological redox reactions. Other bioinorganic electron transport systems include rubredoxins, cytochromes, blue copper proteins, and the structurally related Rieske proteins.

Ferredoxins can be classified according to the nature of their iron-sulphur clusters and by sequence similarity.

Fe_2S_2 *ferredoxins*

Plant-type Ferredoxins: One group of ferredoxins, originally found in chloroplast membranes, has been termed "chloroplast-type" or "plant-type". The active centre is a $[Fe_2S_2]$ cluster, where the iron atoms are tetrahedrally coordinated both by inorganic sulphur atoms and by sulphurs provided by four conserved cysteine (Cys) residues.

In chloroplasts, Fe_2S_2 ferredoxins function as electron carriers in the photosynthetic electron transport chain and as electron donors to various cellular proteins, such as glutamate synthase, nitrate reductase and sulphite reductase. In hydroxylating bacterial dioxygenase systems, they serve as intermediate electron-transfer carriers between reductase flavoproteins and oxygenase.

Adrenodoxin-type Ferredoxins

Adrenodoxin, putidaredoxin and terpredoxin are soluble Fe_2S_2 proteins that act as single electron carriers. In mitochondrial monooxygenase systems, adrenodoxin transfers an electron from NADPH:adrenodoxin reductase to membrane-bound cytochrome P450. In bacteria, putidaredoxin and terpredoxin serve as electron carriers between corresponding NADH-dependent ferredoxin reductases and soluble P450s. The exact functions of other members of this family are not known, although *Escherichia coli* Fdx is shown to be involved in biogenesis of Fe-S clusters. Despite low sequence similarity between adrenodoxin-type and plant-type ferredoxins, the two classes have a similar folding topology.

Thioredoxin-like Ferredoxins

The Fe_2S_2 ferredoxin from *Clostridium pasteurianum* (*Cp*2FeFd) has been recognised as distinct protein family on the basis of its amino acid sequence, spectroscopic properties of its iron-sulphur cluster and the unique ligand swapping ability of two cysteine ligands to the $[Fe_2S_2]$ cluster. Although the physiological role of this ferredoxin remains unclear, a strong and specific intraction of *Cp*2FeFd with the molybdenum-iron protein of nitrogenase has been revealed. Homologous ferredoxins from *Azotobacter vinelandii* and *Aquifex aeolicus* (*Aa*Fd) have been characterised. The crystal structure of *Aa*Fd has been solved. *Aa*Fd exists as a dimer. The structure of *Aa*Fd monomer is different from other Fe_2S_2 ferredoxins. The fold belongs to the $\alpha+\beta$ class, with first four α -strands and two α-helices adopting a variant of the thioredoxin fold.

Fe_4S_4 and Fe_3S_4 Ferredoxins

The $[Fe_4S_4]$ ferredoxins may be further subdivided into low-potential (bacterial-type) and high-potential ferredoxins.

Low and high-potential ferredoxins are related by the following redox scheme:

High-potential Ferredoxins

oxidised$[Fe_4S_4]^{3+} \xrightarrow{\bar{e}}$ reduced$[Fe_4S_4]^{2+}$oxidised $\xrightarrow{\bar{e}}$ $[Fe_4S_4]^{+}$reduced

Low-potential Ferredoxins

The formal oxidation numbers of the iron ions can be [$2Fe^{3+}$, $2Fe^{2+}$] or [$1Fe^{3+}$, $3Fe^{2+}$] in low-potential ferredoxins. The oxidation numbers of the iron ions in high-potential ferredoxins can be [$3Fe^{3+}$, $1Fe^{2+}$] or [$2Fe^{3+}$, $2Fe^{2+}$].

Bacterial-type Ferredoxins

A group of Fe_4S_4 ferredoxins, originally found in bacteria, has been termed "bacterial-type". Bacterial-type ferredoxins may in turn be subdivided into further groups, based on their sequence properties. Most contain at least one conserved domain, including four cysteine residues that bind to a [Fe_4S_4] cluster. In *Pyrococcus furiosus* Fe_4S_4 ferredoxin, one of conserved Cys residues is substituted with aspartic acid.

During the evolution of bacterial-type ferredoxins, intrasequence gene duplication, transposition and fusion events occurred, resulting in the appearance of proteins with multiple iron-sulphur centres. In some bacterial ferredoxins, one of the duplicated domains has lost one or more of the four conserved Cys residues. These domains have either lost their iron-sulphur binding property, or bind to a [Fe_3S_4] cluster instead of a [Fe_4S_4] cluster.

3-D structures are known for a number of monocluster and dicluster bacterial-type ferredoxins. The fold belongs to the $\alpha + \beta$ class, with 2-7 α -helices and four β -strands forming a barrel-like structure, and an extruded loop containing three "proximal" Cys ligands of the iron-sulphur cluster.

High Potential Iron-Sulphur Proteins

High potential iron-sulphur proteins (HiPIPs) form a unique family of Fe_4S_4 ferredoxins that function in anaerobic electron transport chains. Some HiPIPs have a redox potential higher than any other known iron-sulphur protein (e.g., HiPIP from *Rhodopila globiformis* has a redox potential of ca. 450 mV). Several HiPIPs have so far been characterised structurally, their folds belonging to the $\alpha+\beta$ class. As in other bacterial ferredoxins, the $[Fe_4S_4]$ cluster adopts a cubane-like conformation and is ligated to the protein *via* four Cys residues.

Photorespiration

Photorespiration refers to the alternate pathway for production of Glyceraldehyde 3-phosphate (G3P) by Rubisco, the main enzyme of the light-independent reactions of photosynthesis (also known as the Calvin cycle or the Primary Carbon Reduction (PCR) cycle). Although Rubisco favours carbon dioxide to oxygen, (approximately 3-carboxylations per oxygenation), oxygenation of Rubisco occurs frequently, producing a glycolate and a glycerate. This usually occurs when oxygen levels are high; for example, when the stomata (tiny pores on the leaf) are closed to prevent water loss on dry days. It involves three cellular organelles: chloroplasts, peroxisomes, and mitochondria. Photorespiration produces no ATP.

The Oxidative photosynthetic carbon cycle reaction is catalysed by RuBP oxygenase activity:

$$\text{RuBP} + O_2 \longrightarrow \text{Phosphoglycolate} + \text{3-Phosphoglycerate}$$

The phosphoglycolate is salvaged by a series of reactions in the peroxisome, mitochondria, and again in the peroxisome where it is converted into serine and later glycerate. Glycerate reenters the chloroplast and subsequently the Calvin cycle by the same transporter that exports glycolate. A cost of 1 ATP is associated with conversion to 3-phosphoglycerate (PGA) (Phosphorylation), within the chloroplast, which is then free to re-enter the PCR cycle. One carbon dioxide molecule is produced for every 2 molecules of O_2 that are taken up by Rubisco.

Photorespiration is a wasteful process because G3P is created at a reduced rate and higher metabolic cost (2ATP and one NAD(P)H]) compared with RuBP carboxylase activity. G3P produced in the chloroplast is used to create "nearly all" of the food and structures in the plant. While Photorespiratory carbon cycling results in G3P eventually, it also produces waste ammonia that must be detoxified at a substantial cost to the cell in ATP and reducing equivalents.

Role of Photorespiration

Photorespiration is said to be an evolutionary relic. Photorespiration lowers the efficiency of photosynthesis by removing carbon dioxide molecules from the Calvin Cycle. The early atmosphere in which primitive plants originated contained very little oxygen, so it is hypothesised that the early evolution of Rubisco was not influenced by its lack of discrimination between O_2 and carbon dioxide.

Another theory postulates that it may function as a "safety valve", preventing excess NADPH and ATP from reacting with oxygen and producing free radicals, as these can damage the metabolic functions of the cell by subsequent reactions with lipids or metabolites of alternate pathways.

Ways to Minimise Photorespiration (C_4 and CAM Plants)

Since photorespiration requires additional energy from the light reactions of photosynthesis, some plants have mechanisms to reduce uptake of molecular oxygen by Rubisco. They increase the concentration of CO_2 in the leaves so that Rubisco is less likely to produce glycolate through reaction with O_2.

C_4 plants capture carbon dioxide in cells of their mesophyll (using an enzyme called PEP carboxylase), and they release it to the bundle sheath cells (site of carbon dioxide fixation by Rubisco) where oxygen concentration is low. The enzyme PEP carboxylase is also found in other plants such as cacti and succulents who use a mechanism called Crassulacean acid metabolism or CAM in which PEP carboxylase sequesters carbon at night and releases it to the photosynthesising cells during the day. This provides a mechanism for reducing high rates of water loss (transpiration) by stomata during the day.

This ability to avoid photorespiration makes these plants more hardy than other plants in dry conditions where stomata are closed and oxygen concentrations rise. C_4 plants include sugar cane, corn (maize), and sorghum.

Plastid

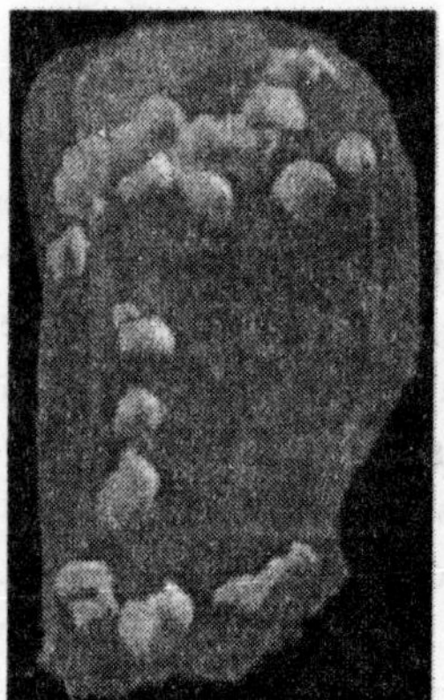

Figure: Plastid

Plastids are major organelles found in plants and algae.

Plastids in Plants

Plastids are responsible for photosynthesis, storage of products like starch and for the synthesis of many classes of molecules such as fatty acids and terpenes which are needed as cellular building blocks and/or for the function of the plant. Depending on their morphology and function, plastids have the ability to differentiate, or redifferentiate, between these and other forms. All plastids are derived from proplastids (formerly "eoplasts", *eo-*: dawn, early), which are present in the meristematic regions of the plant. Proplastids and young chloroplasts commonly divide, but more mature chloroplasts also have this capacity. In plants, plastids may differentiate into several forms, depending upon which function they need to play in the cell. Undifferentiated plastids (*proplastids*) may develop into any of the following plastids:

- Chloroplasts: for photosynthesis;
- Chromoplasts: for pigment synthesis and storage;
- Leucoplasts: for monoterpene synthesis; *leucoplasts sometimes differentiate into more specialised plastids:*
  - — Amyloplasts: for starch storage;
  - — Statoliths: for detecting gravity;
  - — Elaioplasts: for storing fat;
  - — Proteinoplasts: for storing and modifying protein.

Each plastid creates multiple copies of the rectangular 75-250 kilo bases plastid genome. The number of genome copies per plastid is flexible, ranging from more than 1000 in rapidly dividing cells, which generally contain few plastids, to 100 or fewer in mature cells, where plastid divisions has given rise to a large number of plastids. The plastid genome contains about 100 genes encoding ribosomal and transfer ribonucleic acids (rRNAs and tRNAs) as well as proteins involved in photosynthesis and plastid gene transcription and translation. However, these proteins only represent a small fraction of the total protein setup necessary to build and maintain the structure and function of a particular type of plastid. Nuclear genes encode the vast majority of plastid proteins, and the expression of plastid genes and nuclear genes is tightly co-regulated to allow proper development of plastids in relation to cell differentiation.

Plastid DNA exists as large protein-DNA complexes associated with the inner envelope membrane and called 'plastid nucleoids'. Each nucleoid particle may contain more than 10 copies of the plastid DNA. The proplastid contains a single nucleoid located in the centre of the plastid. The developing plastid has many nucleoids, localised at the periphery of the plastid, bound to the inner envelope membrane. During the development of proplastids to chloroplasts, and when plastids convert from one type to another, nucleoids change in morphology, size and location within the organelle. The remodelling of nucleoids is believed to occur by modifications to the composition and abundance of nucleoid proteins.

In plant cells long thin protuberances called stromules sometimes form and extend from the main plastid body into the cytosol and interconnect several plastids. Proteins, and presumably smaller molecules, can move within stromules. Most cultured cells that are relatively large compared to other plant cells have very long and abundant stromules that extend to the cell periphery.

Plastids in Algae

In algae, the term leucoplast (leukoplast) is used for all unpigmented plastids. Their function differ from the leukoplasts in plants. Etioplast, amyloplast and chromoplast are plant-specific and do not occur in algae. Algal plastids may also differ from plant plastids in that they contain pyrenoids.

Inheritance of Plastids

Most plants inherit the plastids from only one parent. Angiosperms generally inherit plastids from the mother, while many gymnosperms inherit plastids from the father. Algae also inherit plastids from only one parent. The plastid DNA of the other parent is thus completely lost.

In normal intraspecific crossings (resulting in normal hybrids of one species), the inheritance of plastid DNA appears to be quite strictly 100 per cent uniparental. In interspecific hybridisations, however, the inheritance of plastids appears to be more erratic. Although plastids inherit mainly maternally in interspecific hybridisations, there are many reports of hybrids of flowering plants that contain plastids of the father.

Origin of Plastids

Plastids are thought to have originated from endosymbiotic cyanobacteria. Due to a split-up into three evolutionary lineages, the plastids are named differently: chloroplasts in green algae and plants, rhodoplasts in red algae and cyanelles in the glaucophytes. The plastids differ by their pigmentation, but also in ultrastructure.

The chloroplasts, e.g. have lost all phycobilisomes, the light harvesting complexes found in cyanobacteria, red algae and glaucophytes, but only in plants and in closely related green algae contain stroma and grana thylakoids. The glaucocystophycean plastid in contrast to the chloroplasts and the rhodoplasts is still surrounded by a remains of the cyanobacterial cell wall. All these primary plastids are surrounded by two membranes.

Complex plastids start by secondary endosymbiosis, when a eukaryote engulfs a red or green alga and retains the algal plastid, which is typically surrounded by more than two membranes, and reduced in its metabolic and/or photosynthetic capacity. Algae with complex plastids derived by secondary endosymbiosis of a red alga include the heterokonts, haptophytes, cryptomonads, and most dinoflagellates (= rhodoplasts). Those that endosymbiosed a green alga include the euglenids and chlorarachniophytes (= chloroplasts). The Apicomplexa, a phylum of obligate parasitic protozoa including the causative agents of malaria (*Plasmodium* spp.), toxoplasmosis (*Toxoplasma gondii*), and many other human or animal diseases also harbour a complex plastid (although this organelle has been lost in some apicomplexans, such as *Cryptosporidium parvum*, which causes cryptosporidiosis). The 'apicoplast' is no longer capable of photosynthesis, but is an essential organelle, and a promising target for anti-parasitic drug development.

Some dinoflagellates take up algae as food and keep the plastid of the digested alga to profit from the photosynthesis; after a while the plastids are also digested. These captured plastids are known as kleptoplastids.

Plastocyanin is an important copper-containing protein involved in electron-transfer. The protein is monomeric, with a molecular weight around 10,500 Da.

Function

In photosynthesis, plastocyanin functions as an electron transfer agent between cytochrome f of the cytochrome b_6f complex from photosystem II and P700+ from photosystem I. Cytochrome b_6f complex and $P700^+$ are both membrane-bound proteins with exposed residues on the lumen-side of the thylakoid membrane of chloroplasts. Cytochrome f acts as an electron donor while P700+ accepts electrons from reduced plastocyanin.

Structure

Although the molecular surface of plastocyanins differs for plants, algae, and cyanobacteria, the structure of the copper binding site is generally conserved. The copper binding site is described as a 'distorted trigonal pyramidal'. The trigonal plane of the pyramidal base is composed of two nitrogen atoms (N_1 and N_2) from separate histidines and a sulphur (S_1) from a ysteine. Sulphur (S_2) from an axial methionine forms the apex. The 'distortion' occurs in the bond lengths between the copper and sulphur ligands. The Cu-S_1 contact is shorter (2.07 Å) than Cu-S_2 (2.82 Å). While the molecular surface of the protein near the copper binding site varies slightly, all plastocyanins have a hydrophobic surface surrounding the exposed histidine of the copper binding site.

In plant plastocyanins, acidic residues are located on either side of a highly conserved tyrosine. Additionally, algal plastocyanins contain similar acidic residues but are shaped differently from those of plant plastocyanins. In cyanobacteria, the distribution of charged residues on the surface is different from eukaryotic plastocyanins and variations among different bacterial species is large. Although the acidic patches are not conserved in bacteria the hydrophobic patch is always present. These hydrophobic and acidic patches are believed to be the recognition/binding sites for the other proteins involved in electron transfer.

Reactions

Plastocyanin ($Cu^{2+}Pc$) is reduced by cytochrome f according to the following reaction:

$$Cu^{2+}Pc + e^{-} \longrightarrow Cu^{+}Pc$$

After dissociation, $Cu^{+}Pc$ diffuses through the lumen until recognition/binding occurs with $P700^{+}$. $P700^{+}$ oxidises $Cu^{+}Pc$ according to the following reaction:

$$Cu^{+} \longrightarrow Cu^{2+}Pc + e^{-}$$

Thylakoid

Thylakoids are a membrane-bound compartment inside chloroplasts and cyanobacteria. They are the site of the light-dependent reactions of photosynthesis. The word "thylakoid" is derived from the Greek *thylakos*, meaning "sac". Thylakoids consists of a thylakoid membrane surrounding a thylakoid lumen. Chloroplast thylakoids frequently form stacks of disks referred to as "grana" (singular: granum). "Grana" is Latin for "stacks of coins". Grana are connected by intergrana or stroma thylakoids, which join granum stacks together as a single functional compartment.

Thylakoid Structure

Thylakoids are membrane-bound structures embedded into the chloroplast stroma.

Membrane

The thylakoid membrane is the site of the light-dependent reactions of photosynthesis with the photosynthetic pigments embedded directly in the membrane. The thylakoid lipid bilayer shares characteristic features with prokaryotic membranes and the inner chloroplast membrane. For example, acidic lipids can be found in thylakoid membranes, cyanobacteria and other photosynthetic bacteria and are involved in the functional integrity of the photosystems. The thylakoid membranes of higher plants are composed primarily of galactolipids that are asymmetrically arranged along and across the membranes. The lipids for the thylakoid membranes are synthesised in a complex pathway involving exchange of lipid precursors between the endoplasmic reticulum and inner membrane of the plastid envelope and transported from the inner membrane to the thylakoids via vesicles.

Lumen

The thylakoid lumen is the compartment bounded by the thylakoid membrane. It plays a vital role for photophosphorylation during photosynthesis. During the light-dependent reaction, protons are pumped across the thylakoid membrane into the lumen making it acidic down to pH 4.

Granum

A granum (plural grana) is a stack of thylakoid discs. Chloroplasts can have from 10 to 100 grana. Grana are connected by stroma thylakoids, also called intergrana thylakoids or lamellae. Grana thylakoids and stroma thylakoids can be distinguished by their different protein composition.

Thylakoid Formation

Chloroplasts develop from proplastids when seedlings emerge from the ground. Thylakoid formation requires light. In the plant embryo and in the absence of light, proplastids develop into etioplasts that contain semi-crystalline membrane structures called prolamellar bodies. When exposed to light, these prolamellar bodies develop into thylakoids. This does not happen in seedlings grown in the dark, which undergo etiolation. An underexposure to light can cause the thylakoids to fail. This causes the chloroplasts to fail resulting in the death of the plant.

Thylakoid formation requires the action of vesicle-inducing protein in plastids 1 (VIPP1). Plants cannot survive without this protein, and reduced VIPP1 levels lead to slower growth and paler plants with reduced ability to photosynthesise. VIPP1 appears to be required for basic thylakoid membrane formation, but not for the assembly of protein complexes of the thylakoid membrane. It is conserved in all organisms containing thylakoids, including cyanobacteria, green algae, such as Chlamydomonas, and higher plants, such as Arabidopsis.

Thylakoid Isolation and Fractionation

Thylakoids can be purified from plant cells using a combination of differential and gradient centrifugation. Disruption of isolated thylakoids, for example by mechanical shearing, releases the lumenal fraction. Peripheral and integral membrane fractions can be extracted from the remaining membrane fraction. Treatment with sodium carbonate (Na_2CO_3) detaches peripheral membrane proteins, whereas treatment with detergents and organic solvents solubilise integral membrane proteins.

Thylakoid proteins

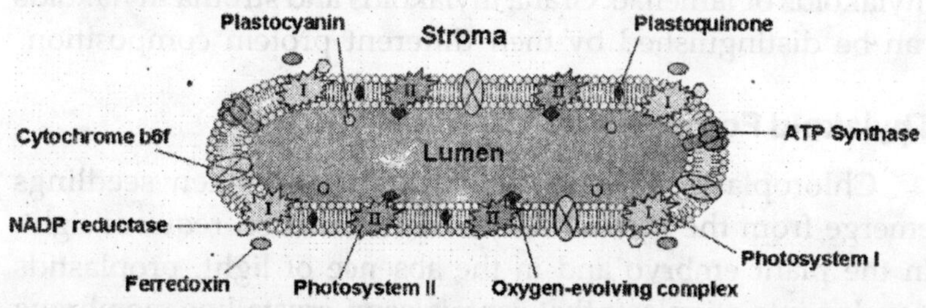

Figure: Thylakoid disc with embedded and associated proteins

Thylakoids contain many integral and peripheral membrane proteins, as well as lumenal proteins. Recent proteomics studies of thylakoid fractions have provided further details on the protein composition of the thylakoids. These data have been summarised in several plastid protein databases that are available online.

According to these studies, the thylakoid proteome consists of at least 335 different proteins. Out of these, 89 are in the lumen, 116 are integral membrane proteins, 62 are peripheral proteins on the stroma side, and 68 peripheral proteins on the lumenal side. Additional low-abundance lumenal proteins can be predicted through computational methods. Of the thylakoid proteins with known functions, 42 per cent are involved in photosynthesis. The next largest functional groups include proteins involved in protein targeting, processing and folding with 11 per cent, oxidative stress response (9 per cent) and translation (8 per cent).

Integral Membrane Proteins

Thylakoid membranes contain integral membrane proteins which play an important role in light harvesting and the light-dependent reactions of photosynthesis. There are four major protein complexes in the thylakoid membrane:

- Photosystems I and II
- Cytochrome b_6f complex
- ATP synthase

Photosystem II is located mostly in the grana thylakoids, whereas photosystem I and ATP synthase are mostly located in the stroma thylakoids and the outer layers of grana. The cytochrome b6f complex is distributed evenly throughout thylakoid membranes. Due to the separate location of the two photosystems in the thylakoid membrane system, mobile electron carriers are required to shuttle electrons between them. These carriers are plastoquinone and plastocyanin. Plastoquinone shuttles electrons from photosystem II to the cytochrome b_6f complex, whereas plastocyanin carries electrons from the cytochrome b6f complex to photosystem I.

Together, these proteins make use of light energy to drive electron transport chains that generate a chemiosmotic potential across the thylakoid membrane and NADPH, a product of the terminal redox reaction. The ATP synthase uses the chemiosmotic potential to make ATP during photophos-phorylation.

Photosystems

These photosystems are light-driven redox centres, each consisting of an antenna complex that uses chlorophylls and accessory photosynthetic pigments such as carotenoids and phycobiliproteins to harvest light at a variety of wavelengths. Each antenna complex has between 250 and 400 pigment molecules and the energy they absorb is shuttled by resonance energy transfer to a specialised chlorophyll *a* at the reaction centre of each photosystem. When either of the two chlorophyll *a* molecules at the reaction centre absorb energy, an electron is excited and transferred to an electron-acceptor molecule. Photosystem I contains a pair of chlorophyll *a* molecules, designated P700, at its reaction centre that maximally absorbs 700 nm light. Photosystem II contains P680 chlorophyll that absorbs 680 nm light best. The P is short for pigment and the number is the specific absorption peak in nanometres for the chlorophyll molecules in each reaction centre.

Cytochrome b_6f Complex

The cytochrome b_6f complex is part of the thylakoid electron transport chain and couples electron transfer to the pumping of protons into the thylakoid lumen. Energetically, it is situated between the two photosystems and transfers electrons from photosystem II-plastoquinone to plastocyanin-photosystem I.

ATP Synthase

The thylakoid ATP synthase is a CF1FO-ATP synthase similar to the mitochondrial ATPase. It is integrated into the thylakoid membrane with the CF1-part sticking into stroma. Thus, ATP synthesis occurs on the stromal side of the thylakoids where the ATP is needed for the light-independent reactions of photosynthesis.

Thylakoid Lumen Proteins

The electron transport protein plastocyanin is present in the lumen and shuttles electrons from the cytochrome b_6f

protein complex to photosystem I. While plastoquinones are lipid-soluble and therefore move within the thylakoid membrane, plastocyanin moves through the thylakoid lumen.

The lumen of the thylakoids is also the site of water oxidation by the oxygen evolving complex associated with the lumenal side of photosystem II.

Lumenal proteins can be predicted computationally based on their targeting signals. In Arabidopsis, out of the predicted lumenal proteins possessing the "TAT" signal, the largest groups with known functions are 19 per cent involved in protein processing (proteolysis and folding), 18 per cent in photosynthesis, 11 per cent in metabolism, and 7 per cent redox carriers and defence.

Thylakoid Protein Expression

Chloroplasts have their own genome, which encodes a number of thylakoid proteins. However, during the course of plastid evolution from their cyanobacterial endosymbiotic ancestors, extensive gene transfer from the chloroplast genome to the cell nucleus took place. This results in the four major thylakoid protein complexes being encoded in part by the chloroplast genome and in part by the nuclear genome. Plants have developed several mechanisms to co-regulate the expression of the different subunits encoded in the two different organelles to assure the proper stoichiometry and assembly of these protein complexes. For example, transcription of nuclear genes encoding parts of the photosynthetic apparatus is regulated by light. Biogenesis, stability and turnover of thylakoid protein complexes is regulated by phosphorylation via redox-sensitive kinases in the thylakoid membranes. The translation rate of chloroplast-encoded proteins is controlled by the presence or absence of assembly partners (control by epistasy of synthesis). This mechanism involves negative feedback through binding of excess protein to the 5 untranslated region of the chloroplast mRNA. Chloroplasts also need to balance the ratios of photosystem I and II for the electron transfer chain. The redox state of the electron carrier plastoquinone in the thylakoid membrane directly affects the transcription of chloroplast genes encoding proteins of the reaction centres of the photosystems, thus counteracting imbalances in the electron transfer chain.

Protein Targeting to the Thylakoids

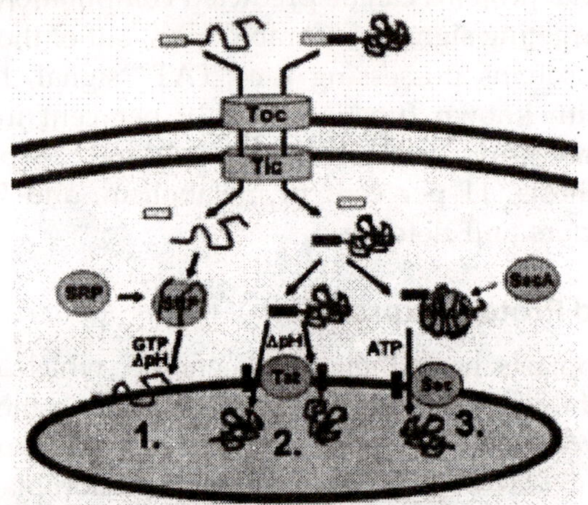

Figure: Schematic representation of thylakoid protein targeting pathways.

Thylakoid proteins are targeted to their destination via signal peptides and prokaryotic-type secretory pathways inside the chloroplast. Most thylakoid proteins encoded by a plant's nuclear genome need two targeting signals for proper localisation: An N-terminal chloroplast targeting peptide, followed by a thylakoid targeting peptide. Proteins are imported through the translocon of outer and inner membrane (Toc and Tic) complexes. After entering the chloroplast, the first targeting peptide is cleaved off by a protease processing imported proteins. This unmasks the second targeting signal and the protein is exported from the stroma into the thylakoid in a second targeting step.

This second step requires the action of protein translocation components of the thylakoids and is energy-dependent. Proteins are inserted into the membrane via the SRP-dependent pathway (1), the Tat-dependent pathway (2), or spontaneously via their transmembrane domains. Lumenal proteins are exported across the thylakoid membrane into the lumen by either the Tat-dependent pathway (2) or the Sec-dependent pathway (3) and released by cleavage from the thylakoid targeting signal. The different pathways utilise different signals and energy sources.

The Sec (secretory) pathway requires ATP as energy source and consists of SecA, which binds to the imported protein, and a Sec membrane complex to shuttle the protein across. Proteins with a twin arginine motif in their thylakoid signal peptide are shuttled through the Tat (twin arginine translocation) pathway, which requires a membrane-bound Tat complex and the pH gradient as an energy source. Some other proteins are inserted into the membrane via the SRP (signal recognition particle) pathway. The chloroplast SRP can interact with its target proteins either post-translationally or co-translationally, thus transporting imported proteins as well as those that are translated inside the chloroplast. The SRP pathway requires GTP and the pH gradient as energy sources. Some transmembrane proteins may also spontaneously insert into the membrane from the stromal side without energy requirement.

Thylakoid Function

The thylakoids are the site of the light-dependent reactions of photosynthesis. These include light-driven water oxidation and oxygen evolution, the pumping of protons across the thylakoid membranes coupled with the electron transport chain of the photosystems and cytochrome b6f complex, and ATP synthesis by the ATP synthase utilising the generated proton gradient.

Water Photolysis

The first step in photosynthesis is the light-driven oxidation (splitting) of water to provide the electrons for the photosynthetic electron transport chains as well as protons for the establishment of a proton gradient.

The water-splitting reaction occurs on the lumenal side of the thylakoid membrane and is driven by the light energy captured by the photosystems.

It is interesting to note that this oxidation of water conveniently produces the waste product O_2 that is vital for cellular respiration. The molecular oxygen formed by the reaction is released into the atmosphere.

Electron Transport Chains

Two different variations of electron transport are used during photosynthesis:

- Non-cyclic electron transport or Non-cyclic photophosphorylation produces NADPH + H^+ and ATP.
- Cyclic electron transport or Cyclic photophosphorylation produces only ATP.

The non-cyclic variety involves the participation of both photosystems, while the cyclic electron flow is dependent on only photosystem I.

- Photosystem I uses light energy to reduce $NADP^+$ to NADPH + H^+, and is active in both non-cyclic and cyclic electron transport. In cyclic mode, the energised electron is passed down a chain that ultimately returns it (in its base state) to the chlorophyll that energised it.
- Photosystem II uses light energy to oxidise water molecules, producing electrons (e^-), protons (H^+), and molecular oxygen (O_2), and is only active in non-cyclic transport. Electrons in this system are not conserved, but are rather continually entering from oxidised $2H_2O$ ($O_2 + 4\ H^+ + 4\ e^-$) and exiting with $NADP^+$ when it is finally reduced to NADPH.

Chemiosmosis

A major function of the thylakoid membrane and its integral photosystems is the establishment of chemiosmotic potential. The carriers in the electron transport chain use some of the electron's energy to actively transport protons from the stroma to the lumen. During photosynthesis, the lumen becomes acidic, as low as pH 4, compared to pH 8 in the stroma. This represents a 10,000 fold concentration gradient for protons across the thylakoid membrane.

Source of Proton Gradient

The protons in the lumen come from three primary sources.

- Photolysis by photosystem II oxidises water to oxygen, protons and electrons in the lumen.
- The transfer of electrons from photosystem II to plastoquinone during non-cyclic electron transport consumes two protons from the stroma. These are released in the lumen when the reduced plastoquinol is oxidised by the cytochrome b6f protein complex on the lumen side of the thylakoid membrane. From the plastoquinone pool, electrons pass through the cytochrome b6f complex. This integral membrane assembly resembles cytochrome bc1.
- The reduction of plastoquinone by ferredoxin during cyclic electron transport also transfers two protons from the stroma to the lumen.

The proton gradient is also caused by the consumption of protons in the stroma to make NADPH from NADP+ at the NADP reductase.

ATP Generation

The molecular mechanism of ATP generation in chloroplasts is similar to that in mitochondria and takes the required energy from the proton motive force (PMF). However, chloroplasts rely more on the chemical potential of the PMF to generate the potential energy required for ATP synthesis. The PMF is the sum of a proton chemical potential (given by the proton concentration gradient) and a transmembrane electrical potential (given by charge separation across the membrane). Compared to the inner membranes of mitochondria, which have a significantly higher membrane potential due to charge separation, thylakoid membranes lack a charge gradient. To compensate for this, the 10,000 fold proton concentration gradient across the thylakoid membrane is much higher compared to a 10 fold gradient across the inner membrane of mitochondria. The resulting chemiosmotic potential between the lumen and stroma is high enough to drive ATP synthesis using the ATP synthase. As the protons travel back down the gradient through channels in ATP synthase, ADP + P_i is combined into ATP. In this manner, the light-dependent reactions are coupled to the synthesis of ATP via the proton gradient.

Transketolase

Transketolase, an enzyme of the Calvin cycle of photosynthesis, catalyses the transfer of a 2-carbon fragment from a 5-carbon keto sugar (D-xyulose-5-P) to a 5-carbon aldo sugar (D-ribose-5-P) to form a 7-carbon keto sugar (sedoheptulose-7-P) and a 3-carbon aldo sugar (glyceraldehyde-3-P).

In mammals, transketolase connects the pentose phosphate pathway to glycolysis, feeding excess sugar phosphates into the main carbohydrate metabolic pathways. Its presence is necessary for the production of NADPH, especially by the brain. Thiamine diphosphate is an essential co-factor, along with calcium.

Diagnostic Use

Red cell transketolase activity is reduced in deficiency of thiamine (vitamin B1), and may be used in the diagnosis of Wernicke's encephalopathy and other B1-deficiency syndromes if the diagnosis is in doubt. Apart from the baseline enzyme activity (which may be normal even in deficiency states), acceleration of enzyme activity after the addition of thiamine pyrophosphate may be diagnostic of thiamine deficiency (0-15 per cent normal, 15-25 per cent deficiency, >25 per cent severe deficiency).

Warburg Effect

The Warburg effect is the inhibition of carbon dioxide fixation, and subsequently photosynthesis, by high oxygen concentrations. The effect is named for Otto Heinrich Warburg. The oxygenase activity of Rubisco, which initiates the process of photorespiration, largely accounts for the Warburg effect. In humans, the Warburg Effect causes cancer cells to rely preferentially on glycolysis for adenosine triphosphate (ATP) production unlike normal healthy cells. Because nicotinamide adenine dinucleotide (NAD) is a co-factor required for glycolysis, by blocking the production of NAD the cancerous cell will eventually die. This allows for the creation of chemotherapy drugs by targeting specific fast growing, NAD reliant cells.

Cellular Metabolism and Fermentation

Glycolysis, the Universal Process

Nine reactions, each catalysed by a specific enzyme, makeup the process we call glycolysis. All organisms have glycolysis occurring in their cytoplasm.

At steps 1 and 3 ATP is converted into ADP, inputting energy into the reaction as well as attaching a phosphate to the glucose. At steps 6 and 9 ADP is converted into the higher energy ATP. At step 5 NAD^+ is converted into NADH + H^+.

The process works on glucose, a 6-C, until step 4 splits the 6-C into two 3-C compounds. Glyceraldehyde phosphate (GAP, also known as phosphoglyceraldehyde, PGAL) is the more readily used of the two. Dihydroxyacetone phosphate can be converted into GAP by the enzyme Isomerase.

The end of the glycolysis process yields two pyruvic acid (3-C) molecules, and a net gain of 2 ATP and two NADH per glucose.

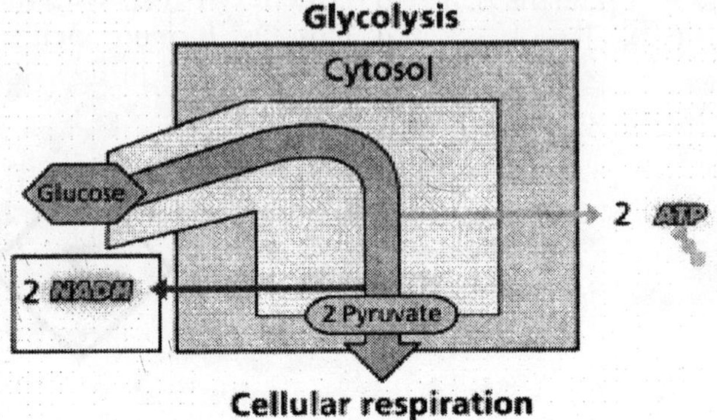

Figure: The glycolysis process

Anaerobic Pathways

Under anaerobic conditions, the absence of oxygen, pyruvic acid can be routed by the organism into one of three pathways: lactic acid fermentation, alcohol fermentation, or cellular (anaerobic) respiration. Humans cannot ferment alcohol in their own bodies, we lack the genetic information to do so. These bio-chemical pathways, with their myriad reactions catalysed by reaction-specific enzymes all under genetic control, are extremely complex. We will only skim the surface at this time and in this course.

Alcohol fermentation is the formation of alcohol from sugar. Yeast, when under anaerobic conditions, convert glucose to pyruvic acid via the glycolysis pathways, then go one step farther, converting pyruvic acid into ethanol, a C-2 compound.

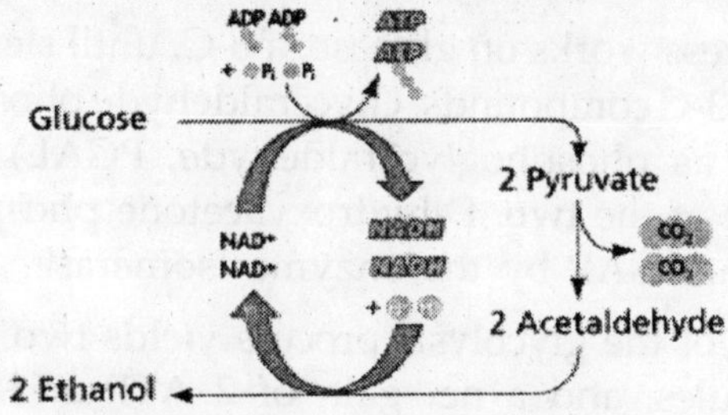

Figure: Fermentation of ethanol

Many organisms will also ferment pyruvic acid into, other chemicals, such as lactic acid. Humans ferment lactic acid in muscles where oxygen becomes depleted, resulting in localised anaerobic conditions.

This lactic acid causes the muscle stiffness couch-potatoes feel after beginning exercise programmes. The stiffness goes away after a few days since the cessation of strenuous activity allows aerobic conditions to return to the muscle, and the lactic acid can be converted into ATP via the normal aerobic respiration pathways.

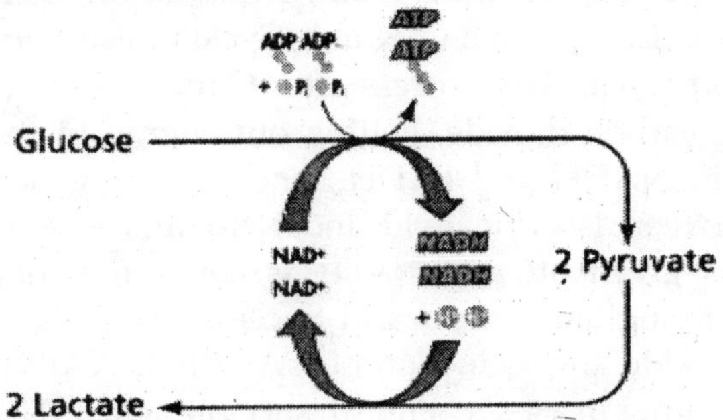

Figure: Fermentation of lactate (lactic acid)

Aerobic Respiration

When oxygen is present (aerobic conditions), most organisms will undergo two more steps, Kreb's Cycle, and Electron Transport, to produce their ATP. In eukaryotes, these processes occur in the mitochondria, while in prokaryotes they occur in the cytoplasm.

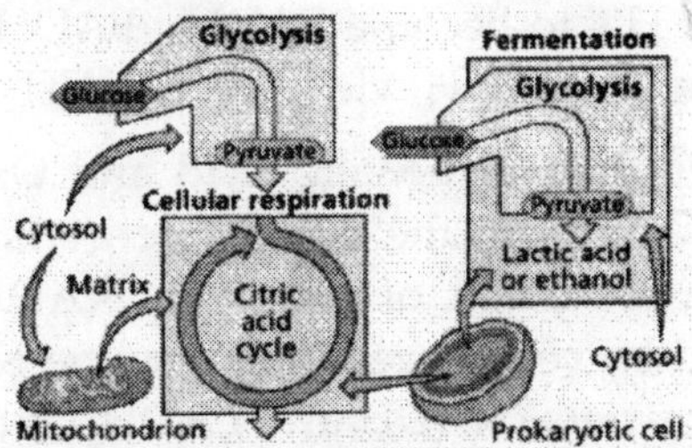

Figure: Overview of the cellular respiration processes

Pyruvic acid is first altered in the transition reaction by removal of a carbon and two oxygens (which form carbon dioxide). When the carbon dioxide is removed, energy is given off, and NAD+ is converted into the higher energy form NADH. Co-enzyme A attaches to the remaining 2-C (acetyl) unit, forming acetyl Co-A. This process is a prelude to the Kreb's Cycle.

Kreb's Cycle (aka Citric Acid Cycle)

The Acetyl Co-A (2-C) is attached to a 4-C chemical (oxaloacetic acid). The Co-A is released and returns to await another pyruvic acid. The 2-C and 4-C make another chemical known as Citric acid, a 6-C. Kreb's Cycle is also known as the Citric Acid Cycle. The process after Citric Acid is essentially removing carbon dioxide, getting out energy in the form of ATP, GTP, NADH and $FADH_2$, and lastly regenerating the cycle. Between Isocitric Acid and Ketoglutaric Acid, carbon dioxide is given off and NAD+ is converted into NADH. Between Ketoglutaric Acid and Succinic Acid the release of carbon dioxide and reduction of NAD^+ into NADH happens again, resulting in a 4-C chemical, succinic acid. GTP (Guanine Triphosphate, which transfers its energy to ATP) is also formed here (GTP is formed by attaching a phosphate to GDP).

The remaining energy carrier-generating steps involve the shifting of atomic arrangements within the 4-C molecules. Between Succinic Acid and Fumaric Acid, the molecular shifting releases not enough energy to make ATP or NADH outright, but instead this energy is captured by a new energy carrier, Flavin adenine dinucleotide (FAD). FAD is reduced by the addition of two H's to become $FADH_2$. $FADH_2$ is not as rich an energy carrier as NADH, yielding less ATP than the latter.

The last step, between Malic Acid and Oxaloacetic Acid reforms OA to complete the cycle. Energy is given off and trapped by the reduction of NAD^+ to NADH. The carbon dioxide released by cells is generated by the Kreb's Cycle, as are the energy carriers (NADH and $FADH_2$) which play a role in the next step.

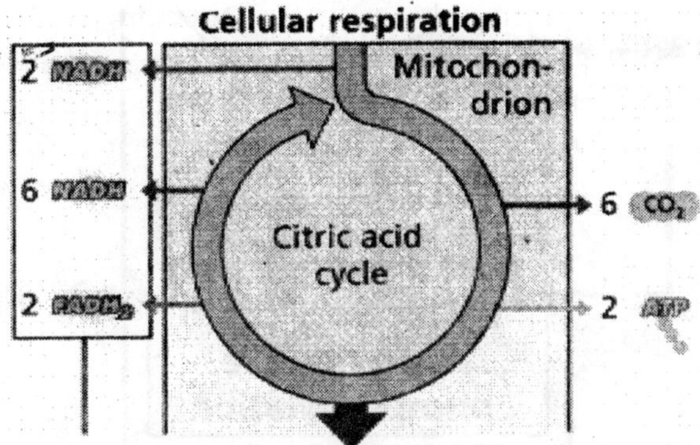

Figure: Summary of the Krebs' (or citric acid) cycle

Electron Transport Phosphorylation

Whereas Kreb's Cycle occurs in the matrix of the mitochondrion, the Electron Transport System (ETS) chemicals are embedded in the membranes known as the cristae. Kreb's cycle completely oxidised the carbons in the pyruvic acids, producing a small amount of ATP, and reducing NAD and FAD into higher energy forms. In the ETS those higher energy forms are cashed in, producing ATP. Cytochromes are molecules that pass the "hot potatoes" (electrons) along the ETS chain. Energy released by the "downhill" passage of electrons is captured as ATP by ADP molecules. The ADP is reduced by the gain of electrons. ATP formed in this way is made by the process of oxidative phosphorylation.

The mechanism for the oxidative phosphorylation process is the gradient of H^+ ions discovered across the inner mitochondrial membrane. This mechanism is known as chemiosmotic coupling. This involves both chemical and transport processes. Drops in the potential energy of electrons moving down the ETS chain occur at three points. These points turn out to be where ADP + P are converted into ATP. Potential energy is captured by ADP and stored in the pyrophosphate bond. NADH enters the ETS chain at the beginning, yielding 3 ATP per NADH. $FADH_2$ enters at Co-Q, producing only 2 ATP per $FADH_2$.

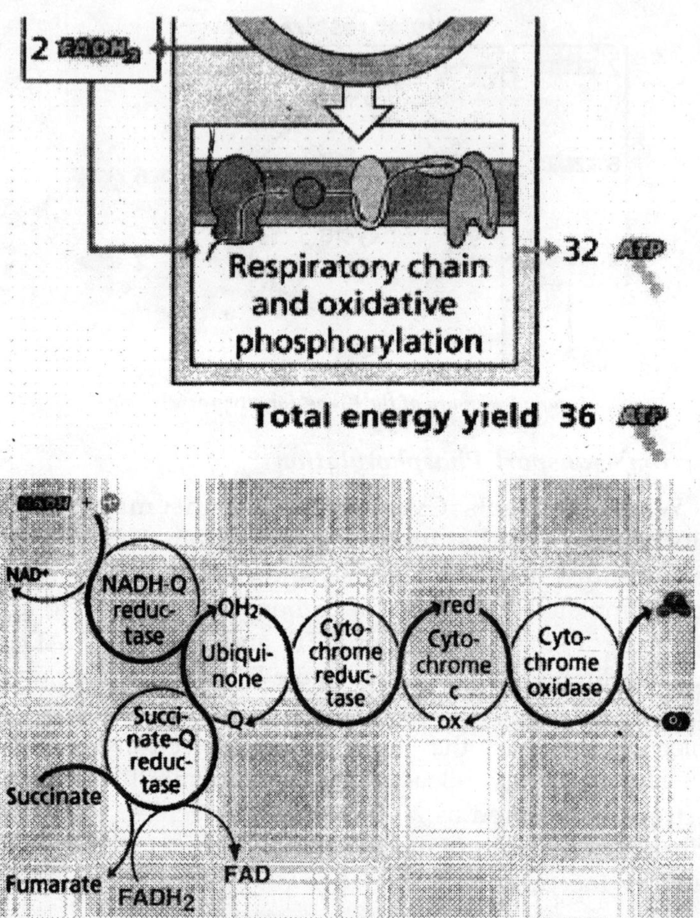

Figure: Electron transport system

The Chemical Logic Behind Fermentation and Respiration

Enzymes relay the electrons released by substrate oxidation to special molecules we call *electron acceptors*. Electron acceptors may be organic or inorganic, and the most common examples thereof are NAD^+ and FAD. Each of these molecules can accept two electrons, yielding $NADH+H^+$ and $FADH_2$, respectively. Since cellular amounts of NAD^+ and FAD are very small, special mechanisms are needed in order to convert $NADH+H^+$

and $FADH_2$ back into NAD^+ and FAD. This is performed through electron transfer from $NADH+H^+$ and $FADH_2$ to other molecules, which may occur through either fermentation or respiration. Contrary to general belief, the distinction between these two processes does not lie on a requirement for O_2.

Fermentation

In fermentation, NADH (or $FADH_2$) donates its electrons to a molecule produced by the same metabolic pathway that produced the electrons carried by NADH (or $FADH_2$). For instance, during intense physical exercise by muscles, NADH generated through glycolysis transfers its electrons to pyruvate (an organic molecule produced by glycolysis), yielding lactate.

H_3C $NADH + H^+$ NAD^+ H_3C H OH

Piruvato **Lactato**

This process is called *lactic fermentation*. Many other kinds of fermentation have been found in microorganisms, and the most well-known among these is alcoholic fermentation:

H_3C H^+ CO_2 H_3C H $NADH + H^+$ NAD^+ H_3C H H OH

Piruvato **Acetaldeido** **Etanol**

Respiration

In respiration, the final acceptor of NADH (or $FADH_2$) electrons is not a product of the metabolic pathway that released the electrons carried by NADH (or $FADH_2$). Many microorganisms use SO_4^{2-}, SeO_4^{2-}, NO_3^-, NO_2^-, NO, U^{6+} (uranium),

Fe^{3+}, H^+, etc. as final electron acceptors. Mammals use O_2, and their respiration is therefore called aerobic respiration. Aerobic respiration happens in the inner mitochondrial membrane, which contains the relevant electron-transferring protein complexes. Each of these complexes accepts electrons from a molecule and transfers them to a different compound, and the full assembly is therefore termed the electron transport chain:

- *NADH Dehydrogenase or Complex I:* In mammals, this complex contains more than twenty polypeptide chains, many of which with no known function. This complex accepts two electrons from NADH+H^+ and transfer them, through Fe-S clusters, to a lipophylic molecule, ubiquinone (Q), which gets converted to ubiquinol (QH_2). In this protein complex, electron transfer releases enough energy to transfer protons (H^+) from the mitochondrial matrix to the intermembrane space, decreasing its pH relative to the matrix.
- *Sucinate Dehydrogenase or Complex II:* This is the sole membrane-bound enzyme in the citric acid cycle. It oxidises succinate to fumarate and transfers both released electrons to FAD, yielding $FADH_2$. As in complex I, these electrons ultimately get transferred to ubiquinone.
- *Cythocrome bc_1 or Complex III:* It accepts electrons from ubiquinol (generated by complexes I & II), and transfers them to cythocrome c, a small soluble protein present in the intermembrane space.
- *Cythocrome c Oxidase or Complex IV:* It receives four electrons from cythocrome c and transfers them to O_2, thereby reducing it to two water molecules.

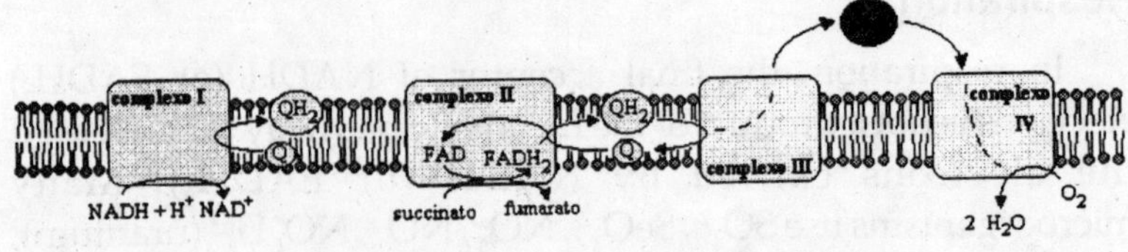

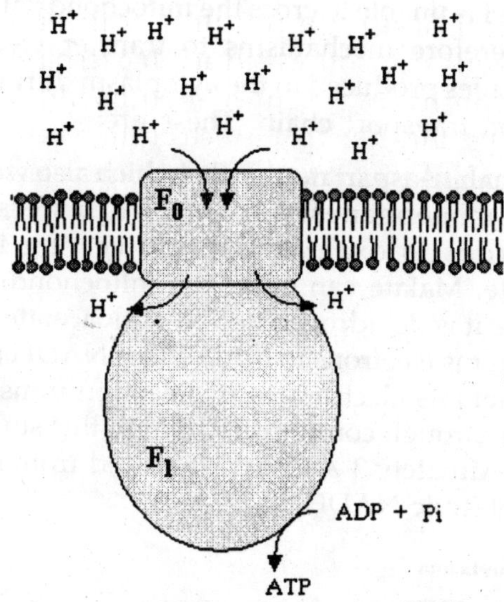

In complexes I, III and IV, electron transfer releases enough energy to transfer H^+ from the mitochondrial matrix to the intermembrane space. This causes an increase of H^+ concentration (and electric potential) in the intermembrane surface, *i.e.* a larger chemical potential of H^+ in the intermembrane space relative to the matrix.

However, when we have two solutions with different concentrations on both sides of a membrane, solute tend to diffuse from the regions of higher chemical potential to areas of lower potential (for a neutral species, this is equivalent to moving from areas of higher concentration to areas of lower concentration). The inner mitochondrial membrane is not permeable to H^+. Under normal conditions, the only way for protons to flow back to the matrix is through a special protein: *ATP synthetase.*

This complex protein contains two major portions: an intermembrane proton channel (F_0) coupled to a catalytic protein complex (F_1) facing the mitochondrial matrix. The F_1 portion contains several subunits with different functions, and converts the energy released by the return of protons to the matrix into chemical energy used to synthesise ATP from ADP and Pi. NADH is unable to cross the mitochondrial membrane. There are therefore mechanisms to transfer electrons from NADH molecules produced in the cytoplasm during glycolysis to the electron transport chain. These are:

> The malate-aspartate shuttle (which also works in gluconeogenesis): NADH transfers its electrons to oxaloacetate, converting it to malate. Malate can enter the mitochondrion, where it is dehidrogenated to oxaloacetate and transfer is electrons to NAD_+. This NADH then transfers its electrons to the electron transport chain through complex I. Through this shuttle, approximately 3 ATP are produced from each cytoplasmic NADH.

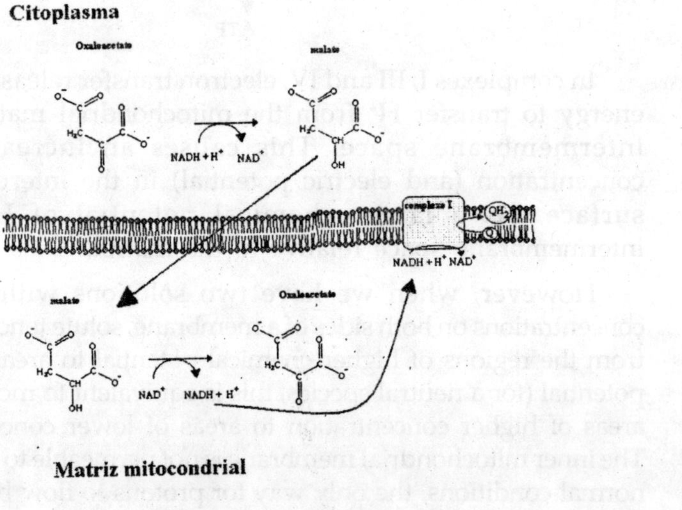

Citoplasma

Oxaloacetato

malato

NADH + H⁺ NAD⁺

complexo I

NADH + H⁺ NAD⁺

malato

Oxaloacetato

NAD⁺ NADH + H⁺

Matriz mitocondrial

Figure: Matriz mitocondrial

> The glycerol-3-P shuttle: In this shuttle, which is very active in brown adipose tissue, cytoplasmic NADH transfer its electrons to the glycolytic intermediate DHAP (dihydroxyacetone phosphate). DHAP is converted to glycerol-3-P, which donates its

electrons to ubiquinone through a FAD-linked glycerol-3-P dehydrogenase located in the outer face of the inner mitochondrial membrane. Through this shuttle, approximately 2 ATP are produced from each cytoplasmic NADH.

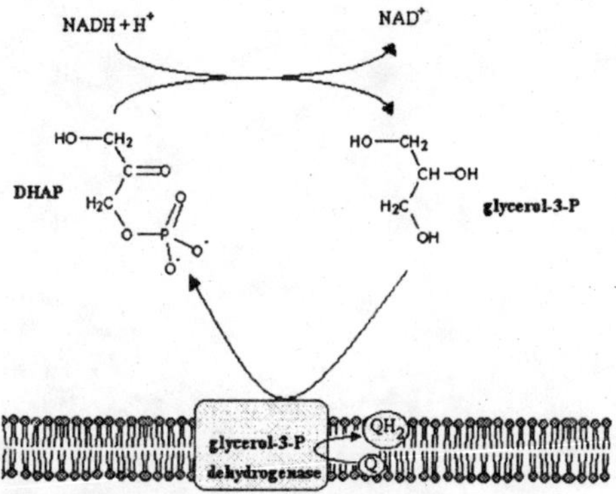

The amount of ATP produced by ATP synthase is therefore related to the difference in H^+ concentration across the membrane. Since NADH oxidation causes proton efflux from the matrix in three protein complexes (I, III & IV), whereas $FADH_2$ oxidation to FAD is only accompanied by such an efflux in two complexes (III & IV), more ATP can be produced from NADH than from $FADH_2$. Oxidation of a NADH molecule produces almost 3 ATP, and $FADH_2$ oxidation yields almost 2 ATP.

Mitochondrial respiration may occur without ATP production, as long as the released protons are able to return to the matrix without passing through the ATP synthetase. This can happen, e.g. if ionophores (lipid-soluble molecules with the ability to transport ions) are added to the mitochondria. In brown adipose tissue, a special protein (thermogenin) forms a proton channel in the mitochondrion inner membrane. The flow of protons back into the matrix through this protein instead of ATP synthetase is responsible for the heat generation characteristic of this tissue.

Converting Processes

The Diversification

Photosynthesis, the Respiratory Chain - Prokaryotes anc Eukaryotes – Diversification

Energy-converting biological processes are usually membrane-bound. In this, they differ fundamentally from simple enzyme reactions happening in solutions. A gain of energy became possible only after membrane-surrounded vesicles, i.e. cells or at least cell precursors had evolved. In the course of the evolution of photosynthesis and the respiratory chain, proton gradients developed. They polarise the membrane, and the energy stored in these gradients is used for the production of an energy-rich bond, a phosphorylation (ADP + P_i > ATP). The cell uses the energy stored in this bond for other processes.

Model experiments show that photophosphorylation, i.e. phosphorylation as a consequence of the use of light energy may also occur under abiotic conditions – though with a small yield only – and in the absence of membranes, but in the presence of molecules that might have existed already shortly after the formation of the earth's surface:

haemin + imidazole + P_i + ADP > *(Light)* > ATP

Primitive prokaryotes can cover their energy expenses by using the electrons of inorganic material (chemolitho-autotrophy), more advanced organisms ferment organic material (heterotrophy).

A simple transformation of light energy into chemical energy occurs in the archaebacterium *Halobacterium halobium*. Just a single protein participates in its 'photosynthesis', while the electron transport chain that is typical for more advanced energy-yielding mechanisms is missing. The energy yield of *Halobacterium* is extraordinarily small.

The photosynthesis that is typical for green plants, blue-green algae, and photosynthetic bacteria is based on the use of light energy (originally with a high portion of UV), i.e. on the conversion of a flow of photons into a flow of electrons and on the existence of electron transport chains.

Photoautotrophic organisms contain usually chlorophyll a or a related derivative like the bacterial chlorophyll *a*. The porphyrin part of chlorophyll molecules contains magnesia ions instead of ferro or feriions that are the central atoms in haemin molecules.

The change to magnesia occurred most likely, because the new complex made better use of the available amount and quality of the light emitted by the sun.

The membrane proteins bound to chlorophyll *a* as well as the composition of the other "accessory pigments" differs among species.

It is assumed that the primitive atmosphere was reducing. Hydrogen, hydrogen sulphide and other sulphide compounds may have acted as electron donators.

Blue-green algae (Cyanophyceae, Cyanobacteria) were the first organisms to use water as both a proton and an electron donator. As a consequence of their early photosynthetic activity, free oxygen entered the atmosphere and its concentration rose from originally almost zero to roughly 20 per cent. This changed the conditions of selection for all organisms considerably. Those able to protect themselves from oxygen and eventually able to use it did finally succeed.

The increase in the amount of oxygen led to the development of an ozone layer in the stratosphere and thus to a shift in the spectral composition of the light reaching the earth surface. Short waves (UV) were now kept back almost completely. The organisms had thus the opportunity to use the visible light increasingly more and as a consequence, the pigment composition of the photosystems changed.

A further understanding of the evolution of energy-converting processes has to consider the evolution of the electron transport chain. Among the similarities of photosynthesis and the respiratory chain is the use of the same proteins: ferredoxin, cytochrome c, cytochrome b, and several others.

The cytochromes b6 (from plastids) and b (from mitochondria) can be homologised due to their topology. Cytochrome b6 corresponds in its change of hydrophobic and hydrophilic amino acid sequences to a sequence at the N-terminus of cytochrome b. The fact that corresponding cytochromes can be homologised is a further proof for a common ancestor of photosynthesis and respiratory chain.

A comparison of the topology of the complexes participating in the electron transport chain leads to a similar conclusion. The respiratory chain is thus a variation of photosynthesis. It was perfected in parallel to the increase of free oxygen.

The photosystems I and II developed at first independently. Combination led to a system, the Z-scheme of photosynthesis, that is far more capable to use light energy efficiently than each of its partial systems alone.

Prokaryotes

Prokaryotes are organisms without a nucleus. Bacteria and blue-green algae (Cyanophyta, Cyanobacteria) belong to them. Prokaryotes were once believed to belong to a uniform organic kingdom, but by now irrefutable evidence that they belong to at least two different kingdoms whose cells differ considerably both in structure and metabolism exist. These two kingdoms are:

- Archaebacteria
- Eubacteria, including blue-green algae.

These two kingdoms have, nevertheless, some things in common, especially the structures required for the maintenance, expression, and the transmission of genetic information. The ribosomes of both kingdoms contain besides a number of proteins ribosomal RNA (rRNA) and tRNA. The ribosomal structure and both types of RNA are among the most conservative supra-molecular complexes found in all cells. The selection pressure working in favour of the maintenance of reliable structures is very high. Yet, even here, modifications occurred in the course of time.

Due to their omnipresence, RNA and ribosomes are especially well-suited as markers in the study of the relations between the different biological kingdoms. Lab techniques make a confinement to the analysis of RNA sensible, and it is not necessary to learn the base sequences in order to understand the degree of relationship and phylogenetic connections. Methods of approximation give just as conclusive results.

Sequence analysis, that can due to the relatively high effort be performed with just a small number of species, may supply additional information. If the sequence of a polynucleotide chain is known, the secondary structure can be determined by optimisation. The secondary structure proved itself to be an even more conservative marker than the presence of certain bases in a given position. Secondary structures remain, even if the nucleotide sequences changed considerably in the course of time, so that homologies can not be determined by comparison of two sequences alone. For the formation of base pairs, an A has not necessarily to have a U opposite itself, a change to a G and a C changes the situation but little. Within a sequence of base pairs, even 'wrong' pairs may occur, because C and A or other base combinations form hydrogen bonds, too, that help stabilising the secondary structure. This argument seems to be inconsistent with what was deduced during the discussion of the evolution of genetic information. The contradiction is nevertheless just a seeming one, as rRNA is a structural component. The base pair sequence contains no genetic information.

A comparison of the rRNA from the chloroplasts of green plants with that of blue-green algae shows, how conservative a secondary structure can be; their structures are highly homologous.

The analysis of rRNA, on the other hand, shows that enormous differences between archaebacteria and eubacteria exist, and that the variability of the rRNA within the archaebaceria is far larger than that within the eubacteria. The combination of all rRNA data shows, that very early during the diversification of the organisms, three separate evolutionary lines evolved that led to the archaebacteria, the eubacteria, and the eukaryotes.

It is assumed, that hypothetical primitive cells, so-called protocells were the common ancestors of all three lines. Prokaryotic cells are with few exceptions like *Mycoplasma* or *Thermoplasma*, surrounded by a cell wall. The chemical composition and the structure of the cell wall is subject to a large variability within the archaebacteria group. The cell walls of eukaryotes, in contrast, are mostly uniform. They consist of one or several layers of murein as is easily shown by Gram's staining method. Due to this reaction, it is distinguished between gram-positive and gram-negative bacteria. This classification turned out to be a reliable taxonomic feature.

The genetic information of prokaryotes is usually stored

in a circular molecule of DNA. In archaebacteria, gene-splicing occurs, which means that the information required for the production of a polypeptide chain is distributed onto several pieces, so-called exons that are separated from each other by non-coding sequences called introns. This type of organisation is not known to occur in eubacteria, too. In some well-studied species, like *Escherichia coli*, gene splicing does not take place meaning that the genome of these species has been optimised during evolution and that all useless genetic sequences have been lost.

The complete nucleotide sequence of strain K12 is known. It contains 4,639,221 base pairs corresponding to 4,288 genes. Thirty eight per cent of these genes have not yet been assigned an essential function within the cell. This again is a hint, that organisms have a far larger supply of genetic information than required for their survival. It seems that the genome is organised in a three-dimensional fashion and that it is well-suited for horizontal gene transfer, since it contains insertion elements and sequences with unusual compositions.

Besides from the single molecule of DNA, a bacterium contains numerous additional small molecules of DNA, the plasmids. They, too, carry genetic information, especially such information that is not always required but may prove life-saving under certain conditions, like, for example, resistance against antibiotics.

Some plasmids are able to integrated into the bacterial chromosome. Also, part of the bacterial chromosome may be cut out and incorporated into a plasmid. Plasmids provide a possibility to transfer genetic information from one cell to another, different species being no insuperable barrier. This process is called horizontal gene transfer. Usually, the consistency of the genome of the respective species remains intact as the foreign DNA entering the cell is recognised and degraded. Bacteria have an extensive array of restriction endonucleases specific for this purpose at their disposal. The bacterial genome is mostly haploid, but the information stored in the plasmids has to be regarded as polyploid, since the cells contain usually several copies of the same type of plasmid.

Beside the exchange of genetic information via plasmid, the cell is able to gain valuable information by taking up DNA. These mechanisms are generally referred to as *parasexuality*. In contrast to the sexuality of eukaryotic cells, where the genetic information of both partners is equal and where both genomes contribute equally to the species' progeny, in parasexuality, the information of one partner is enhanced at the expense of the other's.

All prokaryotes have 70S ribosomes. The protein biosynthesis occurring at these ribosomes is inhibited by chloramphenicol. Bacterial cells are usually very small. It has been calculated that their volume is not large enough to harbour all gene products encoded by the bacterial cell.

Due to a lack of morphological variation, bacteria are classified according to their metabolic performance. The already discussed performances of chemolitoautotropic, heterotrophic, and photoautotrophic bacteria give a good impression of the variety of bacterial metabolisms. Moreover, each cell is able to regulate its metabolic activity according to need. Certain enzymes, especially those participating in the carbohydrate metabolism are only produced when needed, i.e. at sufficient levels of the respective substrate in the medium. Otherwise, the synthesis of these enzymes is inhibited.

This regulation makes for lesser expenses of energy, space, and material. The mechanisms of regulation are extraordinarily variable. They involve the substrate itself, a substrate-specific regulator protein, and a section of DNA that is recognised by the regulator protein. The regulator-DNA complex inhibits the transcription of the subsequent gene or genes and thus also the production of the enzyme (or enzymes) necessary for the catalysis of the respective substrate. These mechanisms show that on a prokaryotic level, processes exist that have to be regarded as differentiations.

Many species form *permanent cells* under unfavourable conditions that are not active metabolically. When the extern circumstances are favourable again, the permanent cells may regenerate into fully functional cells.

Bacteria are usually unicellular. The cells of many species may nevertheless combine in regularly structured aggregates also called *colonies*. Among these aggregates are multicellular filaments and polar trichomes that are screw-like or bundled. Spherical bacteria (cocci) are organised in regular patterns, usually in groups of four. Such colonies are often surrounded by a voluminous jelly coat. Differently structured cells occurring in the same association as is the case with myxobacteria and actinomycetes hint at a division of function. Some aggregations look much like the mycelia of fungi.

Blue-Green Algae

Blue-green algae or *Cyanophyceae* are able to perform a photosynthesis that resembles the green plants very much. The energy-converting processes take place at the outer cell membrane, the plasmalemma and at the intracellular membranes. Intracellular membrane systems do occur in much the same way as in all bacteria, but are not as dominant as in eukaryotic cells. The cell wall consists of murein. The single cells of most species form colonies that are surrounded by often voluminous jelly coats. Blue-green algae are photoautotrophic organisms and have a considerable share in the production of biomass. The ecological importance is emphasised by their dominance in many inshore waters. They are often the cause of "water bloom".

Among the secretions of the cells are, besides the already mentioned jelly coats of varying consistency, also clearly structured, often layered sheaths or capsules that may surround the trichomes. The production of joint sheaths enclosing more than two trichomes lead to a state described as illusory branching. Besides this type, coccal, i.e. non-trichome forming species occur.

In addition, many blue-green algae secrete a whole variety of low molecular metabolites like amino acids, sugars, and others as well as numerous toxic substances.

The trichomes of many species contain at least two functionally different cell types; the bluish green photoautotrophic cells and the usually somewhat more voluminous colourless *heterocysts*. Depending on the species, they occur either terminally or they alternate regularly with the photoautotrophic cells.

Heterocysts are the places where nitrogen fixation is performed. Just as in the case of nitrogen-fixing bacteria, this process is extremely sensitive towards oxygen and very energy-consuming. These features explain the evolution of a diversion of function within a trichome.

The different functional units are separated in space. On the other hand, this separation makes an intercellular transport of materials necessary, otherwise could not be explained.

1. The regular pattern (heterocysts occur at always the same span) and
2. The energy-supply of the heterocysts

In electron-microscopic studies, cell-cell contacts in the form of thin transverse walls with pits emerged. Blue-green algae do often live in association with bacteria. In some cases, certain bacteria are associated with heterocysts. It is assumed that these bacteria consume oxygen thus establishing oxygen-free spaces which helps increasing the rate of nitrogen fixation.

Many blue-green algae form resistant stages like akinetes or *spores*.

Movement by flagella or cilia were found in no species. The typical movement of blue-green algae is a gliding that may set the trichomes into oscillation, the reason why a common genus is called *Oscillatoria*.

About 2000 recent species of blue-green algae have been described. Most are aquatic. Sweet-water species are predominant, marine species are rarer. Some few genera like

Nostoc live on moist soils, others even in dry habitats. Some species live in symbiosis with fungi and form thus new groups of organisms, being either endosymbionts as in the case of *Geosiphon* or living in extracellular associations as in the lichen genera *Peltigera, Collema,* and *Leptogium* or in green plants. A species of *Anabaena,* for example, lives in symbiosis with the water fern *Azolla* and supplies it with nitrogen-containing compounds. *Nostoc* –species are known to live in the thalli of different liverworts, in the root cells of some Cycadean species, and in the rhizomes of *Gunnera.*

Eukaryotes

In comparison to prokaryotes, eukaryotic cells are large. Their surface to volume ratio is less advantageous. They are, moreover, characterised by extensive intracellular membrane systems. The outer membrane (the plasmalemma) is very flexible. It can adapt to extern unevenness, it can form invaginations, and it can give off vesicles either to the inside or to the outside. Eukaryotic cells are with few exceptions aerobic. Compelling evidence for the assumption, that this has not always been the case and that primitive eukaryotic cells have originally been anaerobic exists.

The increase in size required an increase in metabolism. Eukaryotic cells must have developed in environments that were rich in nutriments, since the photosynthetic activity of blue-green algae produced not only large quantities of free oxygen, but also stoecheometric amounts of fixed carbon and made thus for an accumulation of biomass unknown to have existed ever before. Most likely, primitive eukaryotic cells profited of it. Heterotrophic cells used the supply of nutriments and gained size.

How else do eukaryotes differ from prokaryotes?

1. Eukaryotic cells contain certain structure-forming molecular complexes (a cytoskeleton) responsible for movements; actin and myosin, their accompanying regulators, and tubulin.

2. Their genetic information is located on several molecules of DNA (chromosomal equivalents) and is stored in a functional unit enclosed by a membrane, the nucleus.
3. Mitosis, a mechanism for the distribution of chromosomes and chromosome equivalents developed with the help of tubulin.
4. The ability of large and cell wall-less eukaryotic cells to take up organic material led also to their taking up smaller cells, a process called phagocytosis.
5. Phagocytosis may in some cases not have resulted in the digestion of the smaller cells, but in a symbiosis with them, a state also known as endosymbiosis. According to the today mostly accepted endosymbiontic theory, primitive amoeboid eukaryotic cells took up aerobic bacteria as symbionts. Both partners became a unit that was then able to use the decreasing amounts of nutriments far more efficiently. These now aerobic eukaryotes kept their heterotrophic lifestyle.
6. A few of the aerobic eukaryotes started a symbiosis with blue-green algae or prokaryotes similar to blue-green algae and became thus the precursors of plant cells characterised by their photoautotrophic feeding habit.
7. Eukaryotic cells acquired the ability to differentiate, to divide function, and to develop into multicellular organisms.

Eukaryotes can be grouped into four kingdoms:

1. Protista (usually single-celled species)
2. Plants (plantae)
3. Fungi (fungi), and
4. Animals (animalia).

Prokaryotes belong to two further kingdoms, *eubacteria*

and *archaebacteria,* as well as to the evolutionary line leading to the eukaryotes.

There is no doubt that plants fungi and animals developed from ancestors that belonged to the protists. But how did the features listed under (1.) - (7.) came to being? How did their evolution look like?

Actin, myosin and tubulin occur neither in eubacteria nor in archaebacteria. These three proteins cannot be traced back to a common ancestor, since their differences outline their similarities by far.

Actin is one of the most common proteins of animal and many protist cells, of slime moulds, and plant cells. It is a component of contractile elements called microfilaments. It participates in amoeboid movements, and has accordingly a very likely role in phagocytosis. Since amoebas depend on an extern supply of nutriments, the ability to move allowed them to go hunting.

Tubulin is a protein that can just like actin form aggregates with microtubuli, contractile elements, too. It is essential for two types of movements:

1. The movements of chromosomes during mitosis. The nuclear spindle consists mainly of microtubules.
2. The movements of cilia and flagelli. The flagelli and cilia of all eukaryotic cells are organised according to a uniform 9 + 2 scheme. In contrast to the flagelli of bacteria, eukaryotic flagelli and cilia are intracellular structures. The bundles of microtubuli are, just like the rest of the cell's content, surrounded by a continuous membrane, the plasmalemma. Movement of flagelli occurs only, if energy in the form of ATP is supplied.

The nucleus that is surrounded by a double-membrane (a coat), must have developed rather early during evolution as no eukaryotic cells without this type of nucleus are known to exist. The DNA of nearly all eukaryotes is always associated with basic proteins, the histones. The association has a high adaptive value as expressed by the fact that four of the five histones (especially histone IV) have hardly changed during the diversification of plants and animals. Some ciliates, like *Tetrahymena,* contain histone IV that differs from its plant and animal equivalent in 22 per cent of its amino acid residues. This could mean that the interaction between histones and DNA is not yet fully optimised at the evolutionary level of the protists.

The existence of a nucleus is not the same as the existence of a mitosis. In the case of many protists and some fungi, like yeast, no chromosomes can be recognised. During the division of the nucleus and the cell, a seemingly confused tangle of threads appears. The existence of coupling groups seems to have caused a grouping of the genetic material into different structural units. The situation is similar in some chloroplast-containing flagellates like *Euglena.*

In numerous dinoflagellates, several accumulations of fibrillary looking structures occur during the division of both the cell and the nucleus. They could in a functional, but not in a structural sense be called chromosomes. No spindle apparatus is formed. The separation of the genetic material occurs within the nuclear envelope. In other dinoflagellates again, first signs of a spindle apparatus organisation outside of the nucleus have been observed. The spindle fibres, nevertheless, do not make contact with the DNA as the nuclear envelope remains intact and between these two molecular aggregates in these cases, too. In a third group, clearly demarcated areas can be recognised within the membrane in electron-molecular studies that seem to act as specific adaptors between DNA and microtubuli. They may be precursors of the later site of spindle attachment, the *centromer,* but definite proof of homology does not yet exist (D. F. KUBAL, 1975).

In the case of mitosis, what has been said earlier about valuable structures applies here, too: after a mechanism had evolved and was perfected, it remained and was not altered any further. Quite a number of protists, where this state has already been reached, are known. It is striking, that protists contain a large numbers of chromosomes, several hundred actually, while the genetic information of multicellular organisms is concentrated on usually far less than fifty chromosomes. It seems to be simpler to distribute a small number of units regularly and without mistakes onto daughter cells than a large one.

Endosymbiontic Theory

The endosymbiontic theory became more and more likely during the last years as the number of correspondences between mitochondria and chloroplasts became larger and larger the more their details and that of their potential precursors were studied. The probability that these two organelles developed *de novo* in parallel to prokaryotes seems by now to be pretty small. At first glance, the argumentation seems not necessarily conclusive, but the more became known about the organisation and variability of genetic material within cells the easier it became to clarify seeming contradictions.

What is to be said for mitochondria having evolved from bacteria? Mitochondria are semi-autonomously working organelles. They contain DNA and a machinery for protein synthesis. Their ribosomes belong to the 70 S type and chloramphenicol inhibits their protein synthesis. In contrast, the ribosomes found in the cytosol of the eukaryotic host are all of the 80 S type. The host's protein synthesis is inhibited by cycloheximid, but not by chloramphenicol.

The DNA of mitochondria is circular just like in bacteria, though it has to be said that almost all DNA molecules of this size are circular independent of their origin. It is not associated with histones. Its information content is far smaller than that of any species of bacteria. It is assumed that the rest of the information got lost during the course of evolution as it was no longer required for the survival of the symbiont. Meanwhile, it is known that parts of the originally mitochondrial DNA were incorporated into the host's nuclear genome.

Mitochondria have no cell wall. Their inner membrane contains certain lipids that are only found in mitochondria and bacteria. It is assumed that the outer mitochondrial membrane is supplied by the host cell and that it is homologous to a phagocytosis membrane. Its lipid composition does not differ from that of other host cell membranes.

In bacteria and in the inner mitochondrial membrane, the enzymes of the respiratory chain have the same spatial arrangement. The essential enzymes of the bacterial and the mitochondrial respiratory chain are homologous. The highest degree of correspondence exists between the respiratory chain of mitochondria and rhodospirills.

Very few recent eukaryotic cells have no mitochondria. Among them is the amoeba *Pelomyxa palustris.* It lives in symbiosis with aerobic, cell-wall containing bacteria. This is an important evidence for the coming into being of endosymbionts under natural conditions.

Just like mitochondria can be traced back to originally endosymbiontic bacteria, chloroplasts are homologous to blue-green-algae-like prokaryotes. Plastid-containing organisms belong always to the plant kingdom. The position of *Euglena,* a chloroplast-containing group of the euglenophyta is somewhat doubtful as – in contrast to real plants – plastid-free mutants of usually plastid-containing species occur. Even species lacking plastids completely exist. What kinds of similarities exist between chloroplasts and the plastids that stems from them and blue-green algae or their precursors?

1. The degree of homology between the rRNA of chloroplasts and that of recent blue-green algae is very high and is expressed in nearly identical secondary structures. The DNA of chloroplasts, nevertheless, contains introns while that of blue-green algae does not.
2. Chloroplasts and blue-green algae contain the photosystems I and II. The single reactions of the

photosystems of chloroplasts and blue-green algae do mostly correspond, even though the chloroplasts of green algae contain chlorophyll *a* and *b*, while recent blue-green algae contain no chlorophyll *b*. Recent blue-green algae do however contain chlorophyll *a* and phycobilins (phycocyanin and phycoerythrin), a combination that is also known from the plastids of red algae.

3. The large and the small ribosomal subunits of chloroplasts and bacteria complement each other. Reconstitution experiments showed that single ribosomal proteins of bacteria substitute for that of chloroplasts.
4. The DNA of chloroplasts contains just like the DNA of mitochondria relatively little genetic information. Many of the proteins identified and required in chloroplasts are encoded by the nucleus. A comparison between the ATP-synthetases of mitochondria of varying origins and that of chloroplasts shows that the enzymes of both organelles share many similarities. ATP-synthetase consists of several polypeptide chains. Its corresponding genes are partly located in the nucleus, partly in the organelles themselves. The results given as a scheme show that it is not always the same set of genes that is located in the nucleus or the organelles, respectively.
5. In some algae and euglenophyta, so-called cyanelles have been found. Cyanelles are inclusions, intermediates between blue-green algae and chloroplasts. The cyanelles of the euglenophyte *Cyanophora paradoxa*, for example, contain only 5 – 10 per cent of the blue-green algae's genome. Cyanelles are surrounded by a murein-containing wall and live in intracellular vesicles. The genes of the large and the small subunit of ribulose-1.5- bisphosphate carboxylase are located in the cyanelle genome. In green plants, the small subunit is encoded by the nucleus, the

large subunit by the plastid. Cyanelles and host cells depend on each other, none of them can be cultivated alone, i.e. their integration is already far advanced, but has not progressed as far as that of chloroplasts and their host cells.

In 1976, R. A. LEWIN discovered a species called *Prochloron* that is similar to blue-green algae and lives endosymbiontically in the cells of a sponge. It contains chlorophyll *a* and *b* and its thylakoids are stacked in the way that is typical for chloroplasts. This led to the assumption that blue-green algae resembling *Prochloron* were more likely to be the predecessors of chloroplasts than any other recent species. *Prochloron*, nevertheless, is too specialised in many features, so that this assumption had to be discarded for the time being.

In 1986, a Dutch group of researchers (T. Burger-Wiersma, 1991) described a further species similar to *Prochloron* called *Prochlorothrix hollandica*. It was found in a flat, eutrophic lake, the Loosdrechtsee, were it occurred in masses in 1984. The cells are long and their pigment composition resembles that of the chloroplasts of green plants (T.Burger-Wiersma, 1991).

The large differences in the pigmentation and the ultrastructure of the chloroplasts of red algae, brown algae, diatoms, and green plants make a single incident as the initiation of endosymbiosis unlikely. Moreover, it looks as if the single plant groups acquired their plastids separately. The plant kingdom does thus seem to be of polyphyletic origin. The endosymbionts do most likely belong to different groups of blue-green algae.

The ATP-synthetase of bacteria and blue-green algae resembles that of organelles and consists of the same amount of subunits. This finding points at a prokaryotic origin of the enzyme and makes it probable that during evolution, genes from the endosymbionts or the organelles derived from them migrated into the host cell's nucleus and were incorporated into its genome. These results show, too, that the migrations of single genes have to be regarded as independent incidents.

In some coccal algae like *Glaucosphaera vacuolaria,* cyanelles with almost no or completely lacking cell walls occur. Some marine and chloroplast-containing diatoms, the bacillariophyceae, live in a facultative symbiosis with the thread-shaped blue-green algae *Richelia.* The symbiont is integrated into the host cell's plasma. This latter example of an incomplete and reversible integration is good instance for the fact that endosymbioses can occur again and again. The other examples show that the single steps leading to the transformation of a prokaryont into an organelle are separate and independent of each other. Cyanelles are thus good models for the in-detail analysis of the process of integration.

The arguments listed when discussing mitochondria show that a regulated integration of the endosymbionts developed that led finally to both mitochondria and chloroplasts becoming indispensable for their host cells. Namely, no eukaryotic cells without organelles besides the already mentioned exceptions exist.

What effects does the cooperation between the single compartments of the cell – the plastids, mitochondria, cytosol, and the nucleus - have ?

Proteins like the ATP-synthetase that consist of several subunits and the production of which depends on the interaction of different genomes received much attention.

Rubisco, its full name is ribulose-1,5-bisphosphatecarboxylase or fraction-1-protein, is quantitatively the most common protein existing. It consists of small and large subunits, that are encoded by the nucleus (the small subunits) and the chloroplasts (the large subunits) in eukaryots. Rubisco occurs in blue-green algae, too, where all genes co-exist within one genome.

In one type of cyanelle, both genes are located within the cyanelle's genome. Isoelectric focusing, a variation of the electrophoretic separation, reveals an intraspecific variation of the small and the large subunit. The analysis of Rubisco's variability within the genus *Nicotiana* opened up a new possibility to deduce phylogenetic connections between the single species.

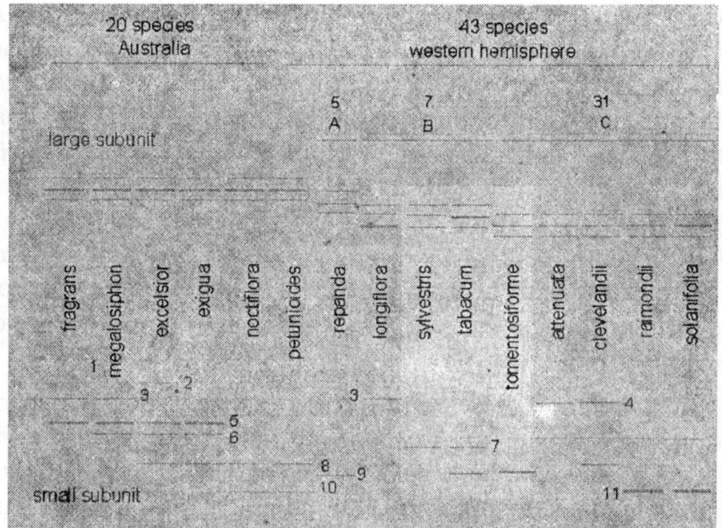

Figure: Schematic Illustration of the Number and the Position of the Small and the Large Rubisco Subunits of Tobacco after their Separation by Isoelectric Focusing.

The Latin names of the *Nicotiana*-species are given (epitheta). It can be seen clearly that the pattern of the large subunit is the same in all Australian species and in some species of the western hemisphere. The rest of the other species gathers into three more groups: A, B, and C. The variability within the small subunit is far larger than that of the large subunit (according to K. Chen; S. Johal, S. G. Wildman, 1976).

The large subunit shows less variation than the small one. This may be caused by the plastids' high degree of polyploidy. Namely, in every cell exist many plastids and each of them has several equal molecules of DNA. The probability that one mutant catches on under these circumstances is far less likely than in a diploid nuclear genome.

The small subunit, too, is usually presented within a gel by more than two bandings. This is quite logical as the gene exists in two different alleles. The cause for this are isoenzymes, i.e. products of different genes that came into being as a consequence of gene duplication or (allo-) polyploidy. The latter is characteristic of many *Nicotiana*-species. *Nicotiana rustica* (n= 24) developed from a hybridisation of *Nicotiana undulata* (n= 12) x *Nicotiana paniculata* (n= 12), while *Nicotiana arentsii* (n= 24) is a hybridisation product of *Nicotiana wigandioides* (n= 12) x *Nicotiana undulata* (n= 12). The hybrids contain the gene products, in this case the small subunit of Rubisco, of both parental species. Both are expressed in equal amounts. The large subunit is encoded by the chloroplast genome and as a consequence, only the genotype of the female parent can be found in the hybrid.

This method allowed to furnish the proof that the species *Nicotiana tabacum* (n= 24) is an amphidiploid hybrid of the maternal species *Nicotiana sylvestris* and the paternal species *Nicotiana tomentosiformis* as it contains the same type of large subunits as *Nicotiana sylvestris*. These subunits can only have been passed on to the new species *Nicotiana tabacum* by the maternal plasma.

An intraspecific variation of Rubisco may overlie its interspecific variation. *Nicotiana suaveolens*, for example, has a highly polymorphic phenotype. The large subunit is the same in all populations while the small one varies. Hybridisation experiments showed that it is an alloenzyme.

Allopolyploidy is far spread in the genus *Nicotiana*, so that a dendrogramm based on the alloenzyme data should not be mixed up with a phylogenetic tree of the genus *Nicotiana*.

An order to safeguard statements about phylogenetic relations, morphological, caryological, and protein biochemical data have to be brought into line. This has already been done successfully in the case of the genera *Brassica*, *Triticum*, and *Solanum*.

Evolution of Species and that of Proteins Follow Different Patterns

Twenty two different types of the glutamat-oxaloacetate-transaminase (GOT) have been identified in European populations of three species of the Genus *Capsella* with the aid of gel electrophoresis. 17 of them were found in *Capsella bursa-pastoris*, 17 in *Capsella grandiflora*, and 19 in *Capsella rubella*. The frequency of six of the Seventeen bandings was smaller than one per cent in all studied populations occurring between Greece and Scandinavia, while the six others had frequencies of more than 10 per cent.

Combinations of different types occur in the single populations. Each pattern represents a certain genotype. All in all, 214 different genotypes were found. Some of them appeared in all three species, while others are typical for one species each. Eighty of the 214 different genotypes, for example, exist only in *Capsella bursa-pastoris*. Geographic differences in the distribution of certain GOT-bandings were identified (H. Hurka, University of Osnabruck, 1983).

This type of analysis shows that the variation of genotypes does not necessarily coincide with species boundaries. The intraspecific variation is just as big as the interspecific variation. These results can be compared to the already discussed situation of *Clarkia speciosa*, where a distribution boundary of two subspecies does not correspond with that of their caryotypes.

Specialisation

Everything seems to point at primitive eukaryotic cells having been amoeboid and able to phagocyte other cells.

Only in rare cases, endosymbiosis or the unification of two cells occurred increasing the probability of a gene exchange, a sexuality. It is considered a well-known fact that a genetic exchange or a recombination of parts of two genomes can only be successful, if both partners are almost or completely identical–the species concept.

A large genetic variability has most likely already existed among the primitive eukaryotic cells leading among other things to the existence of many incompatible genomes. The uptake of one cell by another did therefore usually lead to the lysis of the latter causing the necessity of protection. This may have been the reason behind the different strategies that caused the basically different cell types:

1. Fast and mobile cells, escaping by moving
2. Increase in size and protection by a solid cell wall
3. Settlement in habitats where amoeboid cells cannot survive.

The first strategy led to the evolution of flagellates and ciliates. In comparison to most amoebas, flagellates are relatively small. Movement is an energy-consuming process, and small cells do therefore have an advantage against large ones. Movement occurs usually by one or two flagella. In ciliates, usually the whole cell surface is studded with short cilia that resemble flagella in both structure and function.

The cells are spherical, elongated or spindle-like. Within certain boundaries, the cell shape is flexible, but the degree of variability is far smaller than in amoebae. The position of the cilia marks the cell's polarity. The cilium does often spring from the front pole of the cell, i.e. anterior, and the opposite pole is called posterior.

Many species secrete carbohydrate-containing, jelly-like material at the base of the cilium that can be detected by specific lectines labelled with fluorescence and that do thus constitute a further polarity marker. Some flagellates secrete a building substance like cellulose or chitin at their posterior pole and build thus a specifically structured shell called a lorica that can be glued to solid substances. These flagellates have thus adopted a secondarily sessile lifestyle.

Euglena and species from other systematic groups react phototactically. They contain a light receptor and a mechanism allowing them to orient themselves towards a source of light. Living in light-extensive habitats increases the rate of photosynthesis and thus the energy turnover and the rate of division.

A number of primitive, but partly also of highly developed

recent algae like *Acetabularia, Caulerpa, Vaucheria,* etc. are characterised by an extreme increase in size of single cells, some of them are multinucleate. The stability of their shape is guaranteed by a cell wall. The building substance of the cell walls of most green plants is cellulose, while other polysaccharides, too, occur in some primitive, usually unicellular or few-celled species as well as in non-green algae. In simple, flagellate-like species like *Chlamydomonas* or *Volvox,* the wall consists of several layers and mostly of proteins with a high percentage of carbohydrate. The building molecules are sometimes organised in fibrillary structures.

Cell walls out of cellulose or out of other polysaccharides have in contrast to the rigid cell walls out of murein the advantage of being elastic, since the linear polymers are interconnected 'only' by weak interactions, usually hydrogen bonds, that can easily be broken and reformed. This mechanism is an important pre-condition for the elongation characteristic for plant cells.

With murein in contrast, the monomers are almost exclusively linked by strong covalent bonds. The wall of a cell consists therefore in principle only of a single, closed molecule, a murein saccule, and its binding pattern can only be altered enzymatically. The synthesis of new wall parts occurs immediately after cell division. A later enlargement of the cell is usually impossible.

The development of a wall serves in primitive eukaryotes on one hand as a protection against feeding by amoebas, on the other hand does it put up a resistance against the osmotic pressure of the cell content.

Protozoa without a cell wall do constantly have to pump out the water seeping into the cell under expense of energy. The production of a cell wall is pretty energy consuming, too, but is most likely still more economic as a once-in-a-lifetime investment than the permanent pumping of pulsating vacuoles.

Polarity is a characteristic feature of plant and many animal cells. It is, as you have just seen, also typical for flagellates. The polarity of cells seems to be an important prerequisite for specialisation and differentiation. Polarity can not only be recognised by the elongated shape of a cell, cytological and especially studies performed with the electron microscope show that numerous cell contents are distributed asymmetrically. The intracellular gradient of distribution causes different physiological activities within the single cell sections.

Polarity can be demonstrated visually with the aid of several biochemical methods. One possibility is to furnish proof of the secretion of certain glycoconjugates, i.e. carbohydrate-containing components like polysaccharides, glycoproteins, glycolipids, etc. by localising them at the cell surface with the help of fluorescence-labelled lectines. Many of the glycoconjugates that have been detected this way are secreted only at both or even just at one pole of the cell. The excretory products, usually gelatinous substances, are well-suited for anchoring the cells to a solid surface. It does often, too, serve to combine single cells in colonies.

An anchoring to a solid surface is one of the possibilities allowing survival in running waters. Amoebas cannot settle in such biotopes Sessile cells have thus found an ecological niche that was at first protected from predators.

Multicellular Plants and Division of Labour

Multicellularity and polarity are usually directly linked. Non-directional growth by division is rare in the plant kingdom. In the simplest case, multicellular organisms come in to being, because the daughter cells that exist due to the division of a mother cell stay together. If the single cells differentiate, i.e. they specialise, then it is spoken of a cell colony. The differentiation is morphologically visible, if the single cells differ from each other by features like size, pigment composition, etc.

Often nevertheless, the differences are based purely on complicated biochemical traits that can only be recognised under laboratory conditions or by different behaviours. A differentiation of cells within a colony and within a multicellular organism requires a communication of the cells with each other. One cell, for example, can secrete a certain substance that is recognised by its neighbouring cell as a signal and triggers this cell into changing its physiological activity. Those parts of the cell wall that are now between neighbouring cells loose their contact to the extern environment and can develop new functions like durable cell-cell contacts.

Most primitive, multicellular cells are threadlike. Depending on the species, elongation of the thread, i.e. the filament, occurs either by division of the terminal cell or by division of any cell within the filament, a modus called *intercalary division*. In intercalary division, the nuclear spindle is in parallel to the filament's axis, while the metaphase plate is perpendicular to it. The largest part of the cell wall of each cell remains in contact with the extern environment and only the end walls serve to make contact with the neighbouring cells.

The development of almost sphere-like cell aggregations occurs only in rather few species like *Pandorina, Eudorina, Gonium, Pediastrum*, etc. Their aggregations do rarely contain more than 16 cells. In organisations of higher numbers, the single cell has less and less contact with its extern environment and thus also with the supply of nutriments. As long as no efficient supply network exists – and this developed only in higher plants – compact multicellular colonies have hardly any advantages but quite some heavy disadvantages. Ramified filaments and two-dimensional structures form a kind of intermediary type of organisation between filaments and compact colonies. Both are caused by changing orientations of the spindle of division.

The development of hollow spheres as in the case of the Volvocales is a special case. The cells of one colony exchange information even though they remain largely autonomous. A colony reacts always as a whole with a movement towards light.

Cell-Cell Interactions

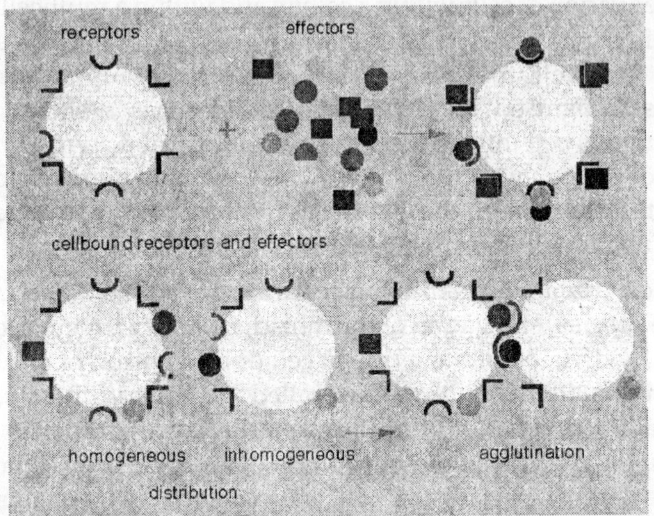

Cooperations between cells that do not stem from the same mother cell require a mutual recognition, i.e. an effector–receptor system has to have evolved. *Effectors* may be soluble and gradient-forming.

Chemotactic substances: substances that are secreted into the medium, but they may also be molecules that are tightly anchored to either the cell membrane or the cell wall.

Signal recipients, i.e. receptors are usually components of the cell surface. After recognition, usually by binding of the effector, a piece of information is passed on to the cell's interior where it is transformed into a physiological reaction. A specific cell-cell interaction is also the pre-condition of the sexual union of two cells. Sexual attractants that are secreted into the medium are called gamones. In the case of some species, their chemical nature has been analysed and some of their structures are given here.

When dealing with algae, you will see that the shape of the sexual cells, the gametes, is characteristic in each systematic group and that the male and female gametes do usually differ considerably, a condition called *anisogamy*. Only in the case of flagellates, flagellate-like species and some algae, the gametes of both sexes look the same, i.e. are isogamous, even though they do often differ in their biochemistry. In these cases, it is spoken of + and – cells or + and – stems or mating types, respectively. Mating occurs only between individuals of a + and a – stem.

Cells without the right type of signal are considered alien. In many species, the ability to mate, i.e. the production of gametes and the respective receptors is triggered by extern factors, like, for example, a lack in the supply of nitrate. The less favourable the living conditions are, the higher is the readiness to mate.

Photosynthesis and Environment

Photosynthesis seems familiar to most everyone, but its function is difficult to understand. Here, the focus is on water used in photosynthesis, and observe the wonderful contrivance of photosynthesis. Water, air, and light exist everywhere on the earth. Plants conduct photosynthesis using those resources and the earth is covered with their green leaves. Because the earth became covered with green plants, animals could have a living environment and were then able to diversify into many species.

Of course, the photosynthetic system with as complicated a mechanism as the present system has, didn't appear suddenly. The mechanism has progressed little by little as evolution progressed. Among water, air, and light, only light has remained unchanged since the beginning of time. And the use of water was the major change that occurred in the photosynthetic mechanism, and it can be said that photosynthesis using water has created a new environment on the earth.

Great Change

Photosynthesis has Changed the Environment on the Earth

When the first organism appeared on the earth, the oxygen concentration in the air seemed to be less than one millionth of what it is now. In an environment like this, the first photosynthesis was conducted using hydrogen sulphide and organic acid in the sea water instead of water itself. But the amount of those materials and the region from which those materials were obtained, were limited. This kind of condition would have led photosynthesis to a standstill. At that time, photosynthesis using water appeared. Through the process of photosynthesis using water, oxygen was produced as a waste. Consequently, the oxygen concentration in the air increased very rapidly. And, as a result of the endless photosynthesis in the plants (algae) using water, Carbon Dioxide, and light, of which were in unlimited amounts, the earth became rich in oxygen, and, the present ecosystem was formed.

Photosynthesis Made the Diversity of Organism Possible

What is the role of water in the process of photosynthesis? It is written in a biology textbook like this, $6CO_2 + 6H_2O \longrightarrow C_6H_{12}O_6 + 6O_2$. From this formula, water's role seems to release oxygen (O) from water into the air (O_2). But that is not true. Things become a little more complicated from this point on, but they are very important, too. Water is also used as an electron feeder. The molecular structure is determined by how atoms such as hydrogen and carbon bind together. The electron works as a clamp to bind atoms. The molecular structure of water is H-O-H, and that of Carbon Dioxide is O=C=O. Water used for photosynthesis provides the electron, which works as a clamp to bind hydrogen to the carbon of Carbon Dioxide to produce sugar. This is the mechanism of producing molecules, which is used in the body.

Energy is needed to receive electrons from water. Organic acid and hydrogen sulphide, which were used for photosynthesis in the early stages, worked as electron feeders, too, and less energy was needed to receive electrons from them than was needed to receive them from water. Therefore, even though water was abundant, it was not easy to use. Photosynthesis overcame this energy problem to use water. Now, what change occurred in photosynthesis? The fact is that light energy was received twice. Light energy received once, was not enough to receive electrons from water and use them to bind hydrogen and carbon. But when light energy was received twice, it was enough. This was the birth of a superb and yet the rather complicated process of photosynthesis, which used this wonderful material, water.

Water Changed the Earth and Brought up the Life

The body of organisms is formed by three almost unlimited resources, which are light, water, and carbon dioxide. Oxygen concentration in the air increased rapidly, and probably many organisms were damaged by oxidisation in the early stage. But after awhile, organisms appeared which could use oxygen effectively, a waste product of photosynthesis, and obtain energy effectively. Because oxygen respiration produces energy far more efficiently than fermentation, almost all organisms use oxygen respiration now. This is the result of the splendid recycling of waste. Photosynthesis uses carbon dioxide and discards oxygen, and respiration uses oxygen and discards carbon dioxide. What a wonderful recycling of waste!

Water has become the resource of photosynthesis, which made it possible for organisms to live in wide areas. Water has also produced life and supports the life of organisms. These events enabled the diversity of organisms to take place, and in addition, oxygen was produced.

It is now possible to say that everything started from water.

Photosynthetic Pigment and Evolution

There are various kinds of photo-environments on the earth, such as the rocky top of the Alps, which receives strong sun rays, places under water at a depth of 150 metres, which receive only 1 per cent of the sun light that the surface of the earth receives, and so on. The organisms with a photosynthetic system exist in almost all these various environments. In order to live in these different environments, plants must have pigment that suits the colour and strength in these environments. In fact, there is a great variety of pigment in the organisms which have a photosynthetic system. To compare those pigments, and to know the evolution of pigment are very important for understanding the evolution of plants. How did plants obtain new pigment?

Study the Diversity of Pigment through the Examination of Molecular System

Among Prokaryote which have a photosynthetic system just as plants growing in soil do, Cyanobacteria (blue-green algae) which have chlorophyll *a* and prokaryotic chlorophyta, which have both chlorophyll *a* and *b*, are known. Chlorophyll *b* is found in plants growing in soil and in eukaryotic chlorophyta, which is the origin of plants, but it is not found in Eukaryotes such as red and brown algae. Where did chlorophyll *b* come from and how did it diversify within those organisms? Since it is known that chlorophyll *b* is made from chlorophyll *a* through two enzyme reactions, Lineage analysis is applied using the CAO gene, which is an enzyme gene related to the above reactions. The results showed that the CAO genes of prokaryotic and eukaryotic chlorophyta, and green plants were not obtained individually, but rather originated from a common ancestor. In other words, the common ancestor of prokaryotic chlorophyta and blue-green algae had both chlorophyll *a* and *b*, and blue-green, red, and brown algae have somehow lost chlorophyll *b* in the evolutional process.

Reproduce the Photosynthetic Evolution in the Laboratory

The combination of photosynthetic pigment and protein form a complex. If the pigment's arrangement or its kind changes, the complex can't use photo-energy. Therefore, it was believed that the arrangement and kind of the pigment in the protein were strictly determined. The two important questions are, how photosynthetic plants predicted new pigment and how they were able to prepare the best suitable protein for pigment in advance. Besides, in many cases, isolated pigment from protein produces active oxygen; therefore, it is a very dangerous molecule to organisms.

These points have made understanding of the process of obtaining pigment, difficult. By the analysis of molecular genealogy, we were able to find the evolutional course of the pigment system. But this fact is not evidence enough to understand, in a practical way, how new pigment is taken into protein and how it functions there. Therefore, CAO gene is put into the algae which didn't have chlorophyll b, and examined what would happen in the cells when chlorophyll *b* entered them for the first time. Isolated chlorophyll *b* is toxic to cells. Algae with the CAO gene would die immediately. But contrary to this expectations, the algae multiplied just like the wild type did. Therefore, it is examined in detail how chlorophyll *b* existed in the cell. The result was this; chlorophyll *b* was taken into the protein which was originally thought not able to bind with chlorophyll *b*, and the protein obtained photo-energy. The protein, then, delivered photo-energy to the nearby pigment, in the usual way. From these findings, evolution may have occurred as follows.

At first, the prokaryon type photosynthetic organism, which had chlorophyll *a*, obtained the CAO gene, and then synthesised chlorophyll *b*. Synthesised chlorophyll *b* was taken into the already existing protein, and then the protein became able to use the light wavelength, which the protein could not originally use, for photosynthesis. Later on, the binding type protein, which exists currently and binds with chlorophyll *b* more practically, was born. As a result of this, chlorophyll *b* was taken into this protein, which became the current pigment style.

This is my scenario of the birth of the current pigment

style. This unexpected scenario was obtained from the experiment of the CAO gene transduction. But this experiment can be applied not only to experiments on the process of obtaining chlorophyll *b* but also to other events, which appeared on the evolutional course of photosynthesis.

Photosynthesis in the Aquatic Environment as Affected by UVR

Since the discovery of the Antarctic ozone 'hole', many studies have been conducted to determine the effects of enhanced UV-B (280-315 nm) on photosynthetic rates of autotrophic organisms. It is accepted now that even natural levels of UVR (280-400 nm) are stressful for some autotrophic organisms. We will summarise what we know about the effects of UVR on photosynthesis of aquatic organisms. Here we consider three major groups–phytoplankton, microphytobenthos (MPB), macroalgae / marine angiosperms – which differ in many ways, especially in regard to their habitats. While phytoplankton live in the water column, MPB and macroalgae occupy the benthic environment. This creates substantial differences with respect to the amount and quality of radiation that they receive.

Thus, although there is a common and general response to UVR of these autotrophic organisms – i.e., inhibition of photosynthesis - there are differences among the groups studied.

These are mainly due to differences in the radiation conditions to which cells are exposed, as well as to the specific sensitivity/acclimation of the organisms under study. To evaluate the overall response of aquatic primary producers to UVR, it is crucial to consider the temporal scale of experimentation, to allow enough time for repair mechanisms and acclimation to UVR. Thus, short–term experiments frequently give an insight about the worst – case scenario for UVR effects on photosynthesis. We also review the effects of UVR upon some related physiological processes (e.g., nutrient incorporation/assimilation, pigment synthesis/bleaching) and morphology (e.g., cell size) that may in turn affect the photosynthetic performance. Finally, to determine the impact of natural and increased levels of UVR upon aquatic ecosystems, we consider the interactive effects of other variables (pH, carbon dioxide concentrations, temperature, etc.) with UVR.

Consequences upon aquatic autotrophic organisms of increased UV-B levels due to ozone depletion events are still uncertain, but changes in biogeochemical cycles, community structure, and trophic web dynamics can be expected.

The photosynthetic process in aquatic ecosystems is responsible for fixing approximately 40 per cent of our planets yearly amount of carbon available for the production of new living matter, with about 48.5 Pg C yr-1 fixed in the aquatic ecosystems (Falkowski 1994, Behrenfeld & Falkowski 1997, Field *et al.* 1998). Carbon fixation in the aquatic environment, mediated by the utilisation of solar radiation, takes place in both the water column and the benthos.

While water-column autotrophic organisms (mainly phytoplankton) are responsible for most of the share in carbon fixation, benthic organisms (i.e., macrophytes and microalgal communities) are involved in about 10 per cent of the total production (Mann & Chapman 1975, Hader *et al.* 1995). Although this latter amount is globally less than that due to phytoplankton, marine macrophytes also provide food (directly or through detritus) to a wide variety of invertebrates and fish in the coastal ecosystems (Duggins *et al.* 1989).

Benthic microalgal communities, on both hard and soft substrata, also serve a crucial ecological function in shallow freshwater and marine habitats. They constitute the local basis of the food webs in shallow areas, which are recognised as having high secondary production (e.g., of fish and their prey). In these areas, the microphytobenthic – MPB - communities may account for 50 per cent or more of the total primary production, equalling or exceeding the productivity of the water column (Underwood & Kromkamp 1999).

Even though solar radiation is attenuated in the water

column (Hargreaves 2003), it penetrates to a depth that will vary, among other things, according to the location (e.g., oceanic vs. coastal), latitude and concentration of particulate and dissolved matter. The euphotic zone in the water column (i.e., 1 per cent of surface PAR, 400-700 nm) can vary from few centimetres in estuarine waters or lakes with a heavy load of DOM (Morris *et al.* 1995, Neale 2001) to more than 100 m in the open ocean (Kirk 1994). Hence ultraviolet radiation (UVR, 280-400 nm) can penetrate accordingly to comparable depths (Smith & Baker 1979, Hargreaves 2003).

In coastal waters, biologically effective ultraviolet B radiation (UV–B, 280-315 nm) reaches only to 1 m depth, as in the Baltic Sea (Piazena & Hader 1997), whereas in the Mediterranean it can penetrate as deep as 20 m (Figueroa 1998). This variability is also observed in other environments: For example, in a study carried out in freshwater Japanese ponds and lakes, Hodoki & Watanabe (1998) determined that the 1 per cent of surface UV-B varied from 0.3 to 2 m, depending mostly on the concentration of chlorophyll-*a* (chl-*a*) and particulate organic carbon present in the water body. The photic zone in the benthic environment extends to ca. 3 mm into the sediment. Fibre optic microsensor measurements have shown that UVR can penetrate down to at least 1.25 mm of this zone, and through scattering it can even exceed the incoming UVR by up to 50 per cent (Bebout & Garcia-Pichel 1995, Garcia-Pichel & Bebout 1996). In addition, as the water column in estuaries and embayments is often shallow, and regularly absent in intertidal areas, UVR can reach high levels at the sediment surface.

Thus, and in view of this background, UVR should be considered a very important environmental factor that can affect different metabolic and physiological processes in autotrophic organisms living in the water column and in the benthos.

We will also discuss the role of UVR in affecting the photosynthetic process in phytoplankton, MPB, and macroalgae. This is especially important as the effects of UVR on the photosynthesis of these organisms may have a considerable impact on higher trophic levels of the aquatic ecosystem (Leech & Johnsen 2003), as well as in climate change.

Methodology to Assess UVR Effects on Photosynthesis

In order to assess UVR effects on photosynthesis, three approaches for exposing algae to UVR are used. These include (1) natural solar radiation, modified by various filters that selectively screen off certain wavebands of radiation; (2) natural solar radiation which is supplemented with artificial UVR from lamps, and (3) fully artificial radiation, implying laboratory experiments. UVR experiments at their best require both that the target organisms are exposed to as realistic a light field as possible, and that high - quality measurements of radiation are obtained. The most realistic results are probably gained from experiments performed under natural solar radiation; artificial radiation sources, however, have also been shown to be very useful for studying mechanistic aspects of UVR responses.

In Situ Incubations

In this type of incubations, the samples are exposed to solar radiation in their natural habitat and at their natural *in situ* depth. To assess the effects of ambient UVR, this approach often involves three types of radiation fields, achieved by filters, i.e. PAR + UV-B + UV-A, PAR + UV-A, and only PAR.

Although *in situ* incubations will result in the most realistic responses, they certainly have the constraint of being conditioned by weather conditions. Therefore, comparatively few in situ studies on the effects of UVR on algal photosynthesis have been conducted, particularly in rough-weather areas, such as the Arctic (Helbling *et al.* 1996b) and Antarctica (Smith *et al.* 1992, Holm-Hansen *et al.* 1993b).

Phytoplankton can be exposed to an *in situ* field of radiation by using UV-transparent bottles hanging from a line or tubes placed in trays which are incubated at different depths in the water column (Smith *et al.* 1992, Holm-Hansen & Helbling 1995). One disadvantage with this approach is that phytoplankton cells are kept at a fixed depth for the entire incubation period (e.g., few hours), thus receiving a constant proportion of the surface incident radiation. In the water column, however, cells are moving within the upper mixed layer (UML) and thus exposed to a variable field of irradiance (Helbling *et al.* 1994, Neale *et al.* 2003). So far, few studies have addressed the importance of mixing rates on the phytoplankton photosynthesis (Helbling *et al.* 1994, Neale *et al.* 1998c, Kohler *et al.* 2001), and with the exception of the experiments performed by Marra (1978) on the effects of PAR, we are not aware of such studies done under *in situ* conditions.

Fixed screens with different filtering capacities have been frequently used to study the *in situ* effect of ambient UVR on shallow - water benthic microalgae in streams (Kiffney *et al.* 1997a), lakes (Vinebrook & Leavitt 1999) and marine habitats (Francoeur & Lowe 1998, Reizopoulou *et al.* 2000). In a four-month *in situ* experiment on UVR effects on MPB communities of a microtidal bay, Wulff *et al.* (1999) used 80 x 80 cm screens placed in wooden frames that were pressed into the sediment. This type of field set-up, however, requires frequent cleaning and careful monitoring of the radiation.

In the case of marine macrophytes (macroalgae and marine angiosperms), most of the *in situ* experimentation has been conducted in the intertidal zone, where access to growing plants is relatively easy (Figueroa *et al.* 1997b, Bischof *et al.* 1998a, Gomez & Figueroa 1998, Flores-Moya *et al.* 1998). Subtidal populations have received less attention due to the complications of working *in situ* at different depths, especially in high latitude zones (Bischof *et al.* 1998a, b). Several authors (Figueroa *et al.* 1997b, Gomez & Figueroa 1998, Flores-Moya *et al.* 1998, 1999) have investigated effects of UVR on macroalgal photosynthesis by incubating algae in their natural environment and monitoring daily variation in photosynthesis and irradiance under different radiation treatments using a similar set up as those described for MPB experiments. More recently, however, efforts have been devoted to analyse in situ photosynthetic activity of subtidal algae.

Considering the practical difficulties of *in situ* incubations, outdoor incubations in temperature controlled containers (e.g., on deck of research vessels, or in flow - through systems on land sites) have been used as an alternative approach. This incubation method is suitable for both short-term (hours) and long-term (days - weeks) experiments carried out with microalgae (Helbling *et al.* 1992b, Holm-Hansen & Lubin 1994, Vernet *et al.* 1994, Sundback *et al.* 1996, Odmark *et al.* 1998), as well as with macroalgae (Hader *et al.* 1996a, b, 1997a, b, Franklin & Forster 1997, Flores-Moya *et al.* 1998, Grobe & Murphy 1998, Altamirano *et al.* 2000).

This set up is often used for determining a worst-case scenario, as samples are exposed to surface (i.e., maximum) incident irradiance. Therefore, neutral density filters are often used to approximately simulate the attenuation of solar radiation in the water column. These filters, however, do not mimic the differential spectral attenuation that actually occurs in the water column (Hargreaves 2003), and samples are generally exposed to higher UV-B / UV-A/ PAR ratios than they would normally experience. It is particularly underwater fluorometer (Beer *et al.* 1998, Ralph *et al.* 1998, Beer & Bjork 2000, Figueroa & Gomez 2001). An alternative approach has been to incubate marine plants for several days at their natural growth site, and after that period the algae were collected and the quantum yield measured using a non-submersible fluorometer (Hanelt *et al.* 1997b, Hader *et al.* 1998, Flores-Moya *et al.* 1999, Figueroa *et al.* 2002).

In contrast to phytoplankton, simulated *in situ* incubations imply fairly realistic light conditions for MPB in the intertidal or littoral zone. This is particularly true when incubating intact sediment cores, as the sediment will provide natural refuges for benthic microalgae, such as motile diatoms and cyanobacteria (Garcia-Pichel & Castenholz 1994, Sundback *et al.* 1996, Odmark *et al.* 1998). Similar approaches have also been used for hard substrata, often involving colonisation of artificial substrata (Bothwell *et al.* 1993).

Supplemented UV-B or UVR

As with several UV experiments carried out with terrestrial organisms (Caldwell *et al.* 1995, Bjorn *et al.* 1999), experimental treatments on aquatic organisms have included the enhancement of ambient UV-B. In some cases, these treatments simulate ozone depletion events. Such experiments, in which natural solar radiation is enhanced by artificial UVR, have been done with phytoplankton (El-Sayed *et al.* 1990, Mostajir *et al.* 1999, Wangberg *et al.* 2001) and microphytobenthos (Sundback *et al.* 1997). A few studies have included simultaneously exclusion and enhancement of UV-B (Odmark *et al.* 1998, Underwood *et al.* 1999). A shortcoming, however, in the majority of experiments using elevated levels of UV-B, has been the use of fixed levels of UV-B for few hours per day. Moreover, the levels of enhancement have varied greatly, from moderate (20 per cent) to ca. 100 per cent above ambient, often resulting in unnatural ratios between PAR and UVR, thus making comparisons between experiments difficult. It is crucial for ecologically relevant studies that the spectral composition of the radiation is realistic (Santas *et al.* 1997). One way to achieve this is to provide additional UV-B so that it mirrors the natural curve, as has been used in terrestrial studies (McLeod 1997). This is possible with a system in which the intensity given is controlled by a computer system linked to a UV-B sensor that continuously measures ambient UV-B levels. This type of set-up allows the simulation of low levels of enhanced UV-B (5 per cent - 20 per cent) as observed during ozone depletion events, and has been used to study the UV-B response of both MPB (Underwood *et al.* 1999, Wulff *et al.* 2000) and phytoplankton (Wangberg *et al.* 2001).

Artificial Radiation

Various artificial radiation sources have been used to assess

UVR effects on aquatic autotrophic organisms. An assorted number of them are commercially available, such as fluorescent and halogen lamps.

So far, most studies carried out with artificial radiation sources have been done with the main objective to determine the impact of UVR at fixed irradiances (Figueroa *et al.* 1997a), or in combination with neutral density screens and cut - off filters to obtain biological weighting functions (BWFs) (Cullen & Neale 1994, Neale & Kieber 2000). In order to determine the sensitivity of intertidal and subtidal algae, Dring *et al.* (1996a, b) used a solar simulator, in which different levels of ozone reduction can be arranged, and Rottgers (1999) had used a similar system to address the response of phytoplankton cultures to changes in UVR. However, it has been found very difficult to mimic the solar radiation spectrum in these types of experiments; in fact very few of the light sources can give reliable results in photobiological research (Brown *et al.* 2000). Moreover, one should be extremely cautious when extrapolating the results obtained in this way to the natural environment.

Materials and Filters

A combination of different materials and filters are normally used to separate different wave- bands of the incident irradiance spectrum. In most of the experiments conducted either in the field or in the laboratory, it is customary to use tubes or vessels made of a material transparent to UVR, such as Quartz, Plexiglas, or Teflon. There are many types of filters that are broadly used in photobiological research, ranging from 'film type' filters, such as Ultraphan, Folex, Mylar-D, and acetate, to 'glass type' filters such as Schott, Hoya and Oriel. Representative spectra of the transmission characteristics of commonly used filters and materials. In general, the materials are long pass filters, and thus they screen off the energy of the lower wavelengths. However, there are filters that allow the energy of just a portion of the spectrum to pass, as is the case of the UG11 filters.

Variables Measured and Experimental Approaches

Various experimental approaches have been used to evaluate the impact of UVR on different cell processes. The evolution of oxygen (Aalderink & Jovin 1997, Hanelt *et al.* 1997a) and incorporation of radiocarbon have been widely used not only to determine the productivity of a water body, but also to assess the impact of UVR. In addition, oxygen microsensors have been shown to be practical tools for high-resolution measurements of UVR effects in sediments and microfilms, particularly in combination with optical microsensors measuring UVR.

In recent years, pulse amplitude modulated (PAM) chlorophyll fluorescence associated with the photosystem II (PSII) has become a useful tool for evaluation of photosynthesis. In fact, chlorophyll fluorescence can function as an indicator of different functional levels in photosynthesis, such as photon capture by light-harvesting pigments, primary light reactions, thylakoid electron transport reactions, dark-enzymatic stroma reactions and slow regulatory feedback processes. The relationship between oxygen evolution and chlorophyll fluorescence in different organisms has also been demonstrated.

Photosynthetic activity has been estimated as chlorophyll fluorescence in macroalgae growing in a variety of water bodies, as in the Arctic, Antarctic, North Sea, Chinese Sea, Mediterranean Sea, tropical or Patagonian. Taking into account the differences in photosynthetic organisation between macroalgae and higher plants, an optimisation of the PAM instrumentation has been needed to meet accurately the low chlorophyll fluorescence emission of macroalgae.

Furthermore, the presence of phycobilisomes in the light-harvesting system of red algae results in generally lower fluorescence values than that measured in green - and brown algae. Due to the increased sensitivity of the PAM fluorescence instrumentation in recent years, this technique has been also used to study UV-B effects on MPB, as well as to address UVR effects on phytoplankton. Studies on the effects of UVR upon phytoplankton have been conducted using both natural communities and monospecific cultures. The exposure of samples has included *in situ* and simulated *in situ* incubations, as well as the use of artificial radiation.

Short-term studies have been generally performed in periods of less than one day, implying that no acclimation is generally allowed, and hence, some of the observed effects represent the worst-case scenario. Still, the majority of UVR studies on phytoplankton photosynthesis have been done using this approach, and they provide a base of comparison among species and different ecosystems. Long-term experiments (i.e., days, weeks), on the other hand, are a preferable choice when making predictions about the effects of UVR on an ecological scale; however, relatively few studies have been performed using this approach.

The response of benthic microalgae to UVR has mainly been assessed by studying natural or semi-natural communities *in situ*, or in outdoor experimental flumes, although laboratory experiments have also been made (Peletier *et al.* 1996, McNamara & Hill 2000). Basically, two types of studies have been conducted: (1) experiments where communities have been allowed to colonise on hard substrata and, (2) experiments where intact natural communities in sediments have been studied.

These two approaches differ in the aspect that the former allows UVR to exert a selective pressure during early growth and succession, which is not the case when studying already established, dense communities with no or little net growth. These two approaches also differ in the choice of target variables. While photosynthetic rate (14C incorporation, oxygen microprofiles), and photochemistry (PAM) have been monitored for MPB in sediments, accrual of biomass has been the most commonly measured variable, particularly in long-term experiments on periphyton on hard substrata. Finally, the impact of UV-B radiation on marine macrophytes has been mostly conducted on individual species and not on the whole community.

The criteria to select species for experimentation / analyses have varied: (a) they are key species due to their contribution in primary production, or because they create a habitat for other marine plants and invertebrates, as the seagrass *Posidonia oceanica* in the Mediterranean Sea, *Laminaria* beds in the North Sea, or *Macrocystis* on the Pacific coast of California (Mann & Chapman 1975), (b) they represent a high share of macroalgal biomass in the ecosystem, as *Ulva* in eutrophic coastal waters (Schramm & Nienhuis 1996) and, (c) they are commercially important as *Porphyra* sp., *Gelidium sequipedale, Macrocystis pyrifera* or *Chondrus crispus*.

Effects of UVR on Phytoplankton Photosynthesis

Short-term Effects: One of the best-known effects of solar radiation upon phytoplanktonic organisms is photoinhibition, which refers to the reduction of photosynthetic rates at relatively high irradiances (Osmond 1994). Many studies have used production – irradiance (P-E) curves to determine phytoplankton photoinhibition due to high PAR levels (Neale & Richerson 1987, Helbling *et al.* 1995); in addition, research has been carried out to determine the additional effects of UVR, not only in tropical but also in temperate and polar regions.

Interestingly, it has been shown that the relative effect of UVR (i.e., as compared to the PAR control) is sometimes higher at lower irradiances. On bright days, when high PAR levels already inhibit the photosystem, UVR produces a relatively lesser effect. This observation, however, depends on many variables, such as the light history of the cells and species composition.

In addition, when phytoplankton cells are exposed to increased levels of solar radiation, they may show a threshold for inhibition, which is followed by a steep increase in photosynthetic inhibition at mid – irradiances, levelling off at higher irradiance values (Helbling *et al.* 1992b, 2001c). However, in some cases no discernible threshold was determined (Behrenfeld *et al.* 1993, Helbling *et al.* 1996b).

In general, when *in situ* incubations are done, UVR causes a sharp decrease in photosynthetic rates (as compared with the PAR – only treatment) especially in surface waters. Even though UV-B radiation is more effective per unit energy (Blumthaler & Webb 2003), and hence potentially more damaging than those at longer wavelengths, many studies conducted in different locations have showed that UV-A is responsible for most of the photosynthetic inhibition, just because their natural levels are much higher (Holm-Hansen *et al.* 1993b, Kim & Watanabe 1994, Villafane *et al.* 1999).

Photosynthesis inhibition decreases with depth, depending, among other things, on water transparency, presence of microorganisms, as well as on incident radiation. The depth distribution of photosynthesis inhibition is highly variable and hence, surface values are not good indicators of the total inhibition in the water column.

Furthermore, when evaluating the integrated photosynthesis inhibition, it is more important to consider the extent of the euphotic zone that is inhibited (e.g., optical depth), rather than the physical depth at which the inhibition is observed. Inhibition of photosynthesis due to UVR is highly variable, depending on the irradiance / doses received by the cells, their specific sensitivity and acclimation potential, as well as the interaction with other variables that can mask the observed effects (mixing, temperature, pH, etc.).

The daily integrated loss of carbon fixation in the euphotic zone in Antarctic waters was calculated to be about 4.9 per cent, under normal ozone column concentrations (Holm-Hansen *et al.* 1993b). At the time of ozone depletion events, which are responsible for a relative increase in incident solar UV-B (Blumthaler & Webb 2003), there was a greater photosynthetic inhibition – reducing daily aquatic primary production by an additional 4-12 per cent (Smith *et al.* 1992, Holm-Hansen *et al.* 1993b).

However, taking into consideration the magnitude and timing of ozone depletion events, the yearly loss of carbon fixation in the Southern Ocean due to these processes was estimated to be < 0.15 per cent (Helbling *et al.* 1994). In addition, some studies (Helbling *et al.* 1994, Neale *et al.* 1998c) have demonstrated that the effects of mixing, i.e., fluctuating radiation regimes (Neale *et al.* 2003), are more important in affecting photosynthesis than the variations in ozone levels. Studies conducted with temperate phytoplankton (Barbieri *et al.* 2002), simulating mixing conditions for Patagonian waters, showed an UVR - induced reduction of photosynthetic rates when the UML extended to a relatively small portion of the euphotic zone.

The use of solar UV-A at low PAR irradiances has also been observed in Californian waters, with a 10-20 per cent increase in photosynthesis due to this effect. For benthic algae, three zones are defined in term of the irradiance received: high, medium and low. While the high irradiance zone is clearly defined as the intertidal zone, the boundaries between medium and low radiation are uncertain and will vary among different water bodies, species considered, etc.

The graphs are examples of UVR – induced photoinhibition in the three groups of algae considered here. In terms of photosynthesis, studies have demonstrated that tropical phytoplanktonic species are more resistant to UVR than those from polar environments (Helbling *et al.* 1992b, 2001c, Villafane *et al.* 1999), probably due to their evolutionary light history with naturally high radiation levels.

In addition, tropical organisms had a higher irradiance threshold for photosynthesis inhibition (Helbling *et al.* 2001c) than polar species (Helbling *et al.* 1992b, Helbling & Villafane 2002), thus providing an additional evidence of their resistance to high UVR levels. Solar radiation increases with altitude (Blumthaler & Rehwald 1992) and thus photosynthesis in lakes located at high altitudes might exhibit enhanced inhibition.

The inhibition of photosynthesis, however, depends not only on the irradiance received at the lake surface, but also on the differences on water temperature, attenuation coefficients and phytoplankton composition among other variables (Neale *et al.* 2001). Biological Weighting Functions (BWFs) (Cullen *et al.* 1992, Neale & Kieber 2000) had also implied the higher resistance of tropical organisms to UVR (Helbling *et al.* 2001c) as compared to those from polar environments (Neale *et al.* 1994, Helbling & Villafane 2002).

Long-term Effects

During long-term experiments, species have the potential to acclimate to new radiation conditions, and processes such as DNA repair and synthesis of photoprotective compounds may occur (Helbling *et al.* 1996a, Holm-Hansen 1997, Roy 2000, Zudaire & Roy 2001, Banaszak 2003).

One of the best ways to test UVR effects on aquatic autotrophic organisms on a long–term basis is by using a "model ecosystem" or mesocosms (Wangberg & Selmer 1997), in which a parcel of the aquatic body is isolated and allowed to progress under similar conditions as in the natural environment.

The main restriction of these experiments is that it is not possible to completely simulate natural conditions – e.g., water movements are restricted and larger organisms are normally excluded. Hence, one should be cautious when interpreting results obtained in these experiments, as other factors (e.g., immigration) are important components when addressing UVR effects from an ecological point of view (Neale *et al.* 2003). Experiments carried out in polar areas (Helbling & Villafane 2002) showed that, at the beginning of experimentation, both Arctic and Antarctic phytoplankton cells were significantly inhibited by UVR.

This inhibition, however, did not increase as the experiment progressed, and growth rates (based either on chl a content or carbon incorporation) were not significantly different between the UVR+PAR and the PAR treatments (Helbling & Villafane 2002). Kim & Watanabe (1994) found that even though short-term exposure to UVR provoked a significant decrease of chl a and photosynthetic rates in two freshwater phytoplankton species, Melosira sp. and Chlorella ellipsoidea, under prolonged UV-A exposure, however, the algae acclimatised by reactivation of the photosystem and enhanced cellular chlorophyll synthesis.

Results from long-term exposure of freshwater phytoplankton are also very variable with no effects determined in an Alpine location (Halac *et al.* 1997), low impact of UVR in a community from a Canadian lake (Laurion *et al.* 1998) and significant changes in phytoplankton composition in a lake from the Andes region (Cabrera *et al.* 1997).

Interactive effects of UVR with other ecological variables are important when addressing photosynthetic inhibition on a long–term basis. In particular, temperature seems to play an important role. For example, the temperate dinoflagellate Prorocentrum micans had a maximum decrease in photosynthetic rates after 21 days of exposure to solar UVR (Lesser 1996), whereas Antarctic phytoplankton had this maximum inhibition after 9 days (Lesser *et al.* 1996).

In addition, research has been conducted to address the interactive effects of UVR and nutrient limitation. There was variability in the responses, with studies that revealed that nutrient– limited cultures were more sensitive to UV-B than nutrient – replete cultures (Cullen & Lesser 1991; Lesser *et al.* 1994); however, Behrenfeld *et al.* (1994) did not find growth inhibition produced by UV-B in nitrogen – limited cultures. Bergeron & Vincent (1997) determined growth rates in different phytoplankton size categories present in a P - enriched system in a Subarctic lake and found different responses according to the wavebands to which cells were exposed.

Effects of UVR on Microphytobenthos Photosynthesis

Although benthic microalgal communities include the same major algal taxa as the pelagic communities, there is a crucial difference. The density of autotrophic, as well as heterotrophic microorganisms is several orders of magnitude higher in benthic communities, resulting in microbial mats or biofilms.

These are characterised by steep physical, chemical and biological gradients, leading to a close spatial and temporal coupling of turnover processes within the mat system (Paerl & Pinckney 1996). Thus, it can be expected that the responses to UVR of these communities are rather complex.

Short-term Effects

Decreased photosynthetic rates (measured as 14 C uptake, oxygen production, or chlorophyll fluorescence) appear to be the most frequently observed short-term effect for MBP, particularly at enhanced UV-B levels.

These results are, however, ecologically relevant only when realistic, moderate increases of UV-levels are used. Wulff *et al.* (2000) found a 50 per cent decrease in 14 C - uptake of MPB on sand when UV-B was increased by 15 per cent above ambient (23 per cent when biologically weighed according to Cullen *et al.* (1992), though only under nutrient depleted conditions.

Using oxygen microsensors, Bebout & Garcia-Pichel (1995) found a dramatic (50-90 per cent) decrease in gross photosynthesis of the surface layers of a cyanobacterial mat (Microcoleus chthonoplastes) under moderate UV-B irradiances (0.35-0.79 W m –2). This decrease was also related to an active downward migration in response to UV-B.

Non-invasive fluorescence measurements on natural diatom biofilms (dominated by Gyrosigma balticum) exposed to supplemented UV-B (7 and 15 per cent above ambient), resulted in a sequence of responses, starting by significantly increased effective quantum yield - O PS II (probably reflecting downwardmigration), followed by a reduction in maximum quantum yield of PS II (F v /F m) and minimal fluorescence (F o) (Underwood *et al.* 1999).

Observed short-term responses of MPB photosynthesis to ambient UV levels are less clear-cut, when compared with supplemented UV-B, and appear to vary with substrate type and community density. For a muddy sediment, no significant effects of ambient UV-B on either carbon uptake or oxygen microprofiles of a diatom mat (dominated by large motile species) were found (Sundback *et al.* 1996).

In a sandy sediment, on the other hand, both carbon uptake and allocation of a community dominated by small (mainly non-motile) diatoms and cyanobacteria, decreased significantly under ambient UV-B, although only at the end of a 3 - week experiment (Odmark *et al.* 1998). Finally, no effects of ambient UV-B on photosynthetic rates were observed in freshwater stream periphyton (Hill *et al.* 1997).

Long-term effects Growth, measured as the accrual of biomass or chl a, is the variable most often studied for long-term UV-B effects on MPB. The clearest negative effects of UV-B on MPB growth have been observed for periphyton colonising artificial substrata.

However, in the majority of these experiments, significant negative effects on growth (30-100 per cent decrease) were only found during the first few weeks. After this UV - inhibition phase, statistically significant negative effects disappeared, or were even reversed.

In one case, the explanation for the reversed effect was the higher sensitivity of grazers than their prey (Bothwell *et al.* 1993, 1994, Sommaruga 2003). When experimenting with already established periphyton communities, however, no detrimental effects were observed (Hill *et al.* 1997, Vinebrook & Leavitt 1998). The effect of UV-B on the growth of natural, established MPB communities inhabiting marine sediments shows a different general pattern than the above-mentioned colonisation experiments.

In sediments, significant effects appear to be fewer, they are more frequently found for rate variables (photosynthesis, C-allocation) than for state variables (biomass, pigment and species composition), and they occur later during the experiment (after 1-2 weeks).

The delay of effects may partly be due to increasing nutrient limitation in the course of the experiments caused by the experimental setup, particularly when working with sandy sediments, which are generally poorer in nutrients than fine sediments.

However, an intriguing question is why the observed effects on rate variables were not reflected more clearly in the state variables? Besides biological reasons, there could be methodological reasons. For example, if a deeper sediment layer than the actual layer affected by UV-B is sampled, there might be a 'dilution' of effects.

Peletier *et al.* (1996) concluded, from laboratory experiments with diatom species isolated from intertidal sediments, that ambient (or even future increased) UV-B is unlikely to affect sediment - inhabiting MPB. Although it may appear that experiments on intact sediment MPB do at least partly support this conclusion (Sundback *et al.* 1996), they do not fully rule out the role of (even ambient) UV-B as a controlling factor (Odmark *et al.* 1998, Underwood *et al.* 1999, Wulff *et al.* 2000).

Odmark *et al.* (1998) found that, while the removal of UV-B created a response, moderate enhancement of UV-B had less obvious effects. This suggests that ambient UV-B can indeed be a factor exerting a selective pressure on MPB particularly in sandy sediments, whereas an increased UV-B exposure due to ozone depletion would not severely affect the type of MPB community studied.

However, given that UV-B, at present level, is a selective force in MPB communities of sandy sediments, there is no a priori reason to assume that the communities should respond to a less degree to an increase in UV-B levels. Moreover, early successional growth phases of sediment communities are indeed, like periphytic communities, susceptible to moderately enhanced (15 per cent) UV-B levels (Wulff *et al.* 2000).

Epipelic communities on sediments of oligotrophic lakes have also shown a significant response to ambient UV-B (Vinebrook & Leavitt 1999).

Are UV-B effects on Microphytobenthos habitat - specific?

The benthic habitats in the above-cited experiments differ in several aspects, such as the type of substratum, water movement and nutrient status. Also the community properties, such as the level of productivity and composition of the food webs (e.g., importance of grazers) vary greatly. Are UVR responses then habitat-specific?

The answer appears to be yes, although similarities in responses also exist. When MPB on both hard and soft substrata (and phytoplankton) were studied simultaneously in alpine oligotrophic lakes in Canada (Vinebrook & Leavitt 1996, 1999) it was found that attached periphyton was affected by ambient UV-B, while epipelon of the sediment and phytoplankton remained unaffected.

On the other hand, Vinebrook & Leavitt (1998) found that ambient UVR had no effect at all on epilithon, while a significant stimulating effect was found for epipelon. This contradicting result was explained by the fact that established epilithon, and not early successional stages, were studied in the latter experiment. An indication of the importance of the substratum type was also found for sediment MPB in a microtidal brackish-water area in Sweden. More variables were affected by UV-B in the sandy than in the muddy sediment (Sundback *et al.* 1996, 1997, Odmark *et al.* 1998).

This suggests that muddy sediments may function as a refuge for MPB (shallow photic zone, dominance of motile diatoms), while in the sandy sediment, UVR penetrates deeper and the MPB is dominated by small - sized attached species. Among community properties, grazing pressure is obviously an important controlling factor for the susceptibility of the MPB to UV-B. Heavily grazed periphyton communities become thin and are thus more sensitive to UV-B (McNamara & Hill 2000). On the other hand, if the grazers are more sensitive to UV-B than the algae, the algal growth will benefit from UV-B (Bothwell *et al.* 1994, Sommaruga 2003).

On a larger geographical scale, climate, latitude / elevation, and general nutrient status of the ecosystem may explain differences in the UV-B responses of MPB. Early colonisation stages of MPB are more susceptible to UV-B than already established, thick communities.

Thus, UV-B can be expected to be a more important controlling factor in for example a cold climate where colonisation events are more frequent due to ice and scouring (Vinebrook & Leavitt 1996). Similarly, freshwater periphyton in mid - latitudes (DeNicola & Hoagland 1996, Hill *et al.* 1997) may be more resistant to current UV-B stress than periphyton communities of higher latitudes (Bothwell *et al.* 1993, Vinebrook & Leavitt 1999). The combined effects of climate warming and increased input of dissolved organic matter (DOM) have been suggested to moderate the effects of UVR - increase in alpine oligotrophic lakes (Vinebrook & Leavitt 1999). However, climate change in combination with acidification may also increase the exposure of organisms to UV-B, particularly in clear, shallow lakes and streams (Schindler *et al.* 1996).

Effects of UVR on Marine Macrophyte Photosynthesis

Studies on comparative primary productivity of marine macrophytes under different scenarios of UV climate are rather scarce; moreover, there is a diffuse picture of their photoadaptive strategies. Taking into account the distinct origin and the morpho - functional divergences of macroalgal species, a common adaptive strategy is unlikely. Thus, a number of responses can be determined among species. However, a general pattern is observed: under natural radiation levels they show daily photoinhibition – a decrease in the photosynthetic rates / yield, at least at high zenith angles. In most cases, high PAR irradiances at noon cause a decrease in photosynthetic rates, but UVR also contributes largely to this process. In intertidal algae in particular, the highest photoinhibition values (mainly due to PAR) are found when low tide coincides with local noon. Even algae harvested from rock pools, where they are normally exposed to extreme solar irradiances, show signs of photoinhibition after prolonged periods of exposure. Under these conditions, increases in temperature and partial dissecation of algal thallus also contribute to the observed photoinhibition.

Deep-water algae and those adapted to shaded environments are inhibited even faster when exposed to direct solar radiation. Recovery of photosynthesis—measured as an increase in fluorescence quantum yield—starts when irradiance begins to decrease, but remains still at saturating levels. Recovery is species—specific and occurs faster in sun - adapted algae than in algae growing at deep or shaded locations and then transferred to the surface. In the eulittoral red algae Porphyra leucosticta (Figueroa *et al.* 1997b), Asparagopsis armata and Felmanophycus rayssae (Jimenez *et al.* 1998) from southern Spain, recovery of photosynthesis occurs immediately after a decrease of only 10 - 20 per cent of solar radiation. However, the brown algae Padina boryana recovers with a 30 per cent irradiance decline, whereas in Sargassum polycystum, a reduction of 70 per cent in the incident radiation is required (Hanelt *et al.* 1994b).

In their review on red macroalgae, Figueroa & Gomez (2001) reported photoinhibition of 30 - 80 per cent at noon, but most of the species showed full recovery in the afternoon. In contrast, only partial recovery was observed in red algae from the North Sea (Dring *et al.* 1996b) or from Patagonian waters (Helbling *et al.*, unpub. data). The recovery of macroalgae after UVR exposure (as compared to the PAR control) is highly variable, with little recovery found in Macrocystis pyrifera (Clendennen *et al.* 1996) and in Gelidium sesquipedale (Gomez & Figueroa 1998), and high recovery with beneficial effects of UV-B in the brown alga Dictyota dichotoma (Flores-Moya *et al.* 1999) and in the marine angiosperm Posidonia oceanica (Figueroa *et al.* 2002). These specific responses provide important information, as the recovery kinetics gives insights into the photoadaptive strategies of macroalgae and their light-stress tolerance capacity.

Thus, those algae capable of dynamic (reversible) photoinhibition under high solar radiation levels and with a rapid recovery capacity will have competitive advantages as compared to those without any efficient photoprotection mechanism. The ability for dynamic photoinhibition during exposure to high radiation, as well as the general degree of photosynthetic adaptation of individual species to different light regimes influences the upper depth distribution of algal zonation. In fact, several taxa and life history stages of inter- and subtidal polar algae show a strong correlation between their depth distribution and their capacity to cope with high radiation stress.

Thus, species growing in the upper subtidal zone show in general more tolerance to high solar radiation levels, especially to UVR, than algae from deeper waters. Relatively few studies have been conducted on a long-term basis to determine the effects of UVR on photosynthesis and growth of macroalgae (Santas *et al.* 1998a, Altamirano *et al.* 2000). In experiments carried out with Ulva sp., UV-B caused a decrease of both growth rates and photosynthesis during the first week of exposure to solar radiation, but UV-A stimulated growth as compared to the PAR treatment (Altamirano *et al.* 2000). However, after two weeks of exposure, no differences were observed between treatments, a fact that hints to the action of acclimation mechanisms, which protect algae against UV stress.

Carbon and Nitrogen Allocation

There is evidence that UVR, especially UV-B, affects carbon allocation in aquatic autotrophic organisms. This has important consequences for food web dynamics, as these changes will affect growth and consequently, the availability of food for other trophic levels, such as bacteria and heterotrophic microorganisms (Sommaruga 2003, Zepp 2003). Changes in lipid, protein, polysaccharide, and fatty acid levels due to UVR have been determined in some phytoplanktonic and MPB organisms.

These studies have especially highlighted the variations in responses, according to the specific sensitivity of the organisms. For example, Buma *et al.* (1996), working with three marine diatoms, found a significant increase in cell protein content when cells were exposed to low UV-B doses, whereas the opposite occurred at higher doses. Veen *et al.* (1997), working with a chlorophyte, demonstrated an increase in cell protein levels when cells were exposed to UVR. Skerrat *et al.* (1998) exposed the diatom Odontella to UV-B radiation and found a reduction in lipid content whereas an increase in Chaetoceros was found. Goes *et al.* (1994, 1997), working with diverse phytoplanktonic species, found changes in the rates and sizes of storage and structural carbohydrates and polyunsaturated fatty acids when exposed to UV-B. Moreover, Dohler (1997a, b) found UVR – induced changes in pool sizes of diverse amino acids of several Antarctic and temperate marine phytoplankton species with UV-A causing, in general, an increase in their levels, whereas UV-B produced the opposite effect.

These results agree with those obtained by Goes *et al.* (1995), who also showed that UV-B caused changes in amino acid concentrations within the cell. Finally, studies performed to determine UVR effects on the ATPcontent of Antarctic phytoplankton, showed a reduction in this component when cells were exposed to UVR (Vosjan *et al.* 1990), but Dohler & Biermann (1994), working with a marine diatom did not find any effect. Studies performed with MPB communities in sediments have also demonstrated changes in carbon allocation as a result of UVR exposure.

The most frequently observed change was a larger relative allocation to proteins at UV-B exposure (Sundback *et al.* 1997, Wulff *et al.* 1999, 2000). This can be interpreted as a larger proportion of fixed carbon spent on growth when carbon fixation decreases, as microalgal cells tend to retain synthesis of proteins rather than storage products under adverse environmental conditions (Maranon *et al.* 1995). Other UV- B effects on carbon allocation were related to polar lipids, which were lower under enhanced UV-B (Odmark *et al.* 1997).

In macroalgae it was also found that UV-B radiation affects carbon and nitrogen allocation, although very few studies have been done in this regard. For example, Altamirano *et al.* (2000) found that more than 78 per cent of the seasonal changes in the internal content of carbon and nitrogen in the green macroalga Ulva rigida were explained by seasonal changes of UV-B.

Mechanisms to Reduce the Effects of UVR on Photosynthesis

Adaptation to UVR assumes the existence of mechanisms that protect the organism or reduce the deleterious effects. According to Roy (2000) four basic mechanisms allow an organism to cope with a stressful situation, i.e., UVR exposure (1) avoidance, (2) reducing the stress by a physiological behavioural mechanism – e.g., through the synthesis of UV - absorbing compounds, (3) repairing the damage produced and, (4) acclimating to the stress allowed enough time. Avoidance mechanisms seem to be a common strategy against exposure to high levels of UVR.

For microalgae living in soft substrata, such as motile cyanobacteria and diatoms, this involves downward vertical migration. Bebout & Garcia-Pichel (1995) showed that by migrating down to 300 μm depth, cyanobacteria could reduce their UV exposure to 10 per cent of that at the sediment surface. For benthic diatoms, the observed downward movement of Gyrosigma balticum at high light levels was first suggested to be related to PAR rather than UV-B. However, a subsequent fluorescence study indicated that the migration could in fact be a direct response to UV-B. Avoidance can also be achieved by means of circadian rhythms that allow an organism to swim down at noon to depths where radiation intensities are low, as occur in some dinoflagellates (Tilzer 1973).

However, it should be observed that UVR can alter the motility and phototaxis of some autotrophic organisms, such as several microalgal species. Moreover, in other organisms, loss of flagella has also been reported. Thus, in some sensitive organisms, avoidance mechanisms can be severely altered by UVR exposure. Another strategy to minimise the effects of UVR is through the presence of UV – screening compounds. The most studied compounds are those collectively named mycosporine like amino acids (MAAs), which are found in many marine and freshwater autotrophic and heterotrophic organisms (Karentz *et al.* 1991b, Dunlap & Shick 1998, Banaszak 2003). Evidence of their protective role upon physiological mechanisms remains still unclear, and in some cases it seems that they just provide partial protection, as in some cyanobacteria (Garcia-Pichel *et al.* 1993). In other cases, though, MAAs have been proved to be an effective protection mechanism (Helbling *et al.* 1996a, Neale *et al.* 1998a) so that photosynthesis in phytoplanktonic cells with higher amounts of MAAs was less inhibited.

In benthic diatoms, however, the production of such protective substances does not appear to be a major strategy. Although MAAs have been detected in MPB of shallow - water subtidal sediments, the concentrations are low and show no significant increase at UV exposure (ambient or increased), which agrees with the findings of Peletier *et al.* (1996). Jeffrey *et al.* (1999) tested 152 algal species and found that diatoms generally had low concentrations of UV - screening compounds as compared with other algal groups. Moreover, Helbling *et al.* (1996a) found that pennate diatoms (which usually dominate benthic diatom communities) contained less MAAs than centric diatoms. Other compounds may also have a protective role, functioning as UV - screening agents (Banaszak 2003).

For example, scytonemin is a UV - absorbing extracellular substance found in the sheath of cyanobacterial filaments (Garcia-Pichel *et al.* 1992). In addition, high concentrations of carotenoids as a result of UVR exposure have been observed in diatom mats (Underwood 1999), some cyanobacteria and chlorophytes (Goes *et al.* 1994), suggesting an UV-protecting function of these pigments. MAAs have also been reported in green, red and brown macroalgae from tropical, temperate and polar regions.

The concentration of MAAs in macroalgae has been found to be related to depth zonation and UV exposure (Karentz *et al.* 1991b). Their accumulation seems to be higher under high than under low daily irradiance values (i.e. different latitudes), and moreover, generally higher in intertidal than that in subtidal algae (Karsten *et al.* 1998b, Karsten & Wiencke 1999, Hoyer *et al.* 2001). In addition, this accumulation seems to be a wavelength-dependent process (Karsten *et al.* 1998a, Franklin *et al.* 1999, 2001, Karsten & Wiencke 1999), and an UV-B-mediated increment of these compounds has been shown in a variety of algae (DeNicola & Hoagland 1996, Molina & Montecino 1996). In Chondrus crispus, both UV-A and UV-B stimulated a strong accumulation of shinorine, whereas the content of palythinol and palythine was mainly stimulated by PAR, indicating a MAA- specific induction triggered by these wavelengths (Karsten *et al.* 1998a).

In Palmaria palmata, on the other hand, and when exposed only to PAR, a 6 - fold increase in the porphyra - 334 concentration was observed; the treatment receiving PAR+UVAgave similar results plus to an accumulation of shinorine; under full solar radiation, accumulation of porphyra - 334, shinorine and palythine was observed (Karsten & Wiencke 1999). In addition, in Chondrus crispus, pre-exposure to blue light followed by growth under natural UV-A led to a 7-fold increase in the synthesis of shinorine as compared with growth without the blue light pre-treatment (Franklin *et al.* 2001).

So, it has been hypothesised that there are two photoreceptors for MAAs synthesis in C. crispus, one for blue light and one for UV-A, which act synergistically (Franklin *et al.* 2001). In macroalgae, other types of potentially protective compounds are also found, such as phlorotanins in brown algae (Pavia *et al.* 1997) and coumarins in the green algae Dasycladus (Gomez *et al.* 1998, Perez-Rodriguez *et al.* 1998).

In addition, and while UVR – mediated DNA damage occurs in aquatic autotrophic organisms (Buma *et al.* 1996, 1997, 2001b, Mitchell & Karentz 1993, van de Poll *et al.* 2001), repair mechanisms of the DNA molecule (Buma *et al.* 2003) are also present (Mitchell & Karentz 1993). However, the presence of one or other mechanism (i.e., photoreactivation, nucleotide excision repair or recombination repair) is clearly dependant on the species under study and the radiation conditions at which the cells are exposed (Buma *et al.* 2003). Finally, acclimation mechanisms to cope with high UVR intensities are important in several aquatic organisms.

These usually occur on a long–term basis, when organisms have been exposed for enough time to the stress factor (UVR). One of these acclimation mechanisms is the previously mentioned synthesis of MAAs, as found in some natural populations and cultures of phytoplankton (Villafane *et al.* 1995b, Helbling *et al.* 1996a, Zudaire & Roy 2001).

However, the synthesis of UV - absorbing compounds is not a general response, and several species do not show an increase of MAAs content even after several weeks of exposition to UVR (Behrenfeld *et al.* 1992, Villafane *et al.* 2000). Acclimation can also occur through a change in the community composition (Villafane *et al.* 1995b), so that those species more adapted to a particular light regime will dominate.

For example, a natural Antarctic phytoplankton population dominated by flagellates (80 per cent in terms of carbon biomass), changed to a diatom dominated population when receiving UVR + PAR; whereas in those samples receiving only PAR, small flagellates still dominated. However, diverse responses are observed in different sites.

For example, Mousseau *et al.* (2000) in their study conducted with an estuarine community also observed changes in diversity when samples were exposed to different radiation treatments. A shift from a diatom – dominated community to small flagellates occurred more rapidly in the treatment receiving enhanced UV-B as compared to those receiving natural UV-B levels. Clearly, responses are strongly species – specific and depend on radiation levels and quality to which organisms are exposed.

Other Photosynthesis - Related Effects

There are a number of UVR effects that are closely related to the photosynthetic performance of aquatic primary producers. These effects are due to the couplings between radiation – especially UVR - and a number of morphological and biochemical factors within the cell (Hessen *et al.* 1997). Thus, for example, radiation – induced changes in nutrient uptake, synthesis and allocation of metabolic compounds, motility / orientation, and cell morphology will result in variations in photosynthetic rates. In the following paragraphs we will outline some of these effects – nutrient incorporation and enzyme activities related to carbon and nitrogen metabolism, accumulation or damage on pigments.

Specific effects, such as DNA damage, which may induce a reduction in growth rates (Buma *et al.* 1995) and hence affecting overall primary productivity, are addressed in Buma *et al.* (2003). Nutrient incorporation/assimilation and enzyme activities growth of aquatic autotrophic organisms is dependent not only on carbon assimilation, but also on the incorporation and assimilation of nitrogen, phosphate, sulphur and several micronutrients (Lobban & Harrison 1997). In general, it is considered that UVR – especially UV-B - is an inhibitor of uptake processes (especially nitrogenous), whereas UV-A stimulates or exerts no significant effects on the uptake of these ions (Dohler 1997a, b). In particular, studies carried out with phytoplanktonic organisms have demonstrated that nitrate and ammonium uptakes are affected by UVR (Dohler 1987, 1992, Dohler & Buchmann 1995).

Furthermore, Dohler (1992, 1995, 1997a, b), working with several Antarctic and North Sea phytoplanktonic species has pointed out the diversity of responses among the organisms tested. Thus, samples dominated by the prymnesiophyte Pheaocystis pouchetii were very sensitive to UV-B doses (in terms of 15 N-ammonium uptake), and so were those containing Ceratium sp., Coscinodiscus sp. and Noctiluca sp. 15N- nitrate uptake was not or only slightly affected by

UV-B irradiances (Dohler 1992). On the other hand, in experiments conducted with North Sea natural phytoplankton populations, Dohler & Hagmeier (1997) found that UV-A radiation stimulated 15N – ammonium uptake. Fewer studies have addressed the effects of UVR on P- uptake of phytoplanktonic cells.

Hessen *et al.* (1995), working with the chlorophyte Chlamydomonas reinhardtii, found a stimulation under low UV-B doses (< 3.6 kJoules m –2 at 312 nm), but higher inhibition when UV-B doses were higher. In addition, studies on UV-B effects carried out in both sandy and muddy sediments have suggested that the nutrient availability may be an important factor for the susceptibility of MPB communities to UV-B exposure (Odmark *et al.* 1998). Wulff *et al.* (2000) designed an experiment to test this hypothesis, and showed that the availability of nutrients indeed can act to mitigate the effects of UV-B on a microbenthic community on a sandy substratum. Some studies have also addressed the UVR effects on nutrient incorporation in marine macroalgae (Franklin & Forster 1997, Hader & Figueroa 1997).

In particular, these studies have focused on UVR effects upon carbon anhydrase (CA) and nitrate reductase (NR) activities. These are important enzymes involved in the incorporation of carbon and nitrogen within the cell (Turpin 1991), thus any stress factor that affects them will ultimately influence photosynthesis. Studies carried out with algae collected from southern Spain found daily variations (i.e., circadian rhythms) in NR and CA activities.

In Dasycladus vermicularis it was found that these variations were antagonistic during the onset of solar radiation, although these changes only partially matched those of photosynthesis (Gomez *et al.* 1998), suggesting that these processes are affected differentially by UVR. In long-term studies, it has been shown that UV-A radiation stimulated NR activity, and UV-B decreased both nitrate uptake and NR activities. On the other hand, UV-B radiation seems to stimulate

CA activity in eulittoral algae but not in subtidal (Figueroa 1998, Vinegla *et al.* 2000b). In addition, experiments were conducted to determine the effects of UVR on the activity of Calvin cycle enzymes, such as ribulose-1,5-biphosphate carboxilase/ oxygenase (Rubisco) and glyceraldehyde -3–phosphate dehydrogenase (G3PDH), and in Arctic macroalgae it was found that the photosynthetic activity decreased due to the negative effects of UVR upon these enzymes (Bischof *et al.* 2000).

Pigments

Several researchers have reported the decrease of photosynthetic pigments due to exposure to UVR. This reduction can be due to a combination of factors, such as the inhibition of de novo synthesis and the natural turnover of pigments, or directly to photobleaching. Bleaching can occur not only by UVR, but also due to exposure to high PAR intensities (Maske & Latasa 1997); it is species – specific and also depends on the spectral characteristics of the radiation treatments imposed to the cells. Helbling *et al.* (1993), working with several marine phytoplankton cultures, found that Nannochloris oculata (Eustigmatophyceae) had a decrease in chl a content of 30, 60 and 80 per cent under PAR only, PAR + UV-A, and PAR + UVR, respectively, after being exposed for 4.5 h to solar radiation.

The prymnesiophyte Isochrysis galbana, on the other hand, did not experience significant changes in chl a content (for the same radiation treatments) even after 7 h of exposure. Other experiments have also demonstrated the differential sensitivity to UVR of various pigments (Quesada *et al.* 1995), with the phycobiliproteins being especially sensitive to these wavelengths (Sinha *et al.* 1995). Absolute amounts of photosynthetic pigments, commonly used as an estimator of growth in autotrophic organisms, seem to be also affected by UVR.

During a long–term experiment, Helbling *et al.* (1992b) simulated ozone depletion events by moving Antarctic phytoplankton towards the Equator, so that the samples were exposed not only to increased levels of UVR, but also to natural changes in the relative proportions of UV-B and UV-A. They found a decrease in the growth rate of Antarctic phytoplankton exposed to UVR as compared to that exposed to the PAR– only treatment. However, growth rates were not significantly different when the samples were incubated under UVR levels similar to those found at their sampling site in the Antarctic. Data on long-term experiments conducted in both polar areas (Helbling & Villafane 2002) showed that even though photosynthesis was initially affected by UVR (day 1), growth rates, evaluated either as carbon fixation or chl a content did not show any significant differences. In general, studies have demonstrated that different growth responses due to UVR exposure occur not only among taxa (Jokiel & York 1984, Villafane *et al.* 2000), but also within the same genus.

For example, in the chlorophyte Dunaliella salina growth rate was not affected by UVR, whereas in D. tertiolecta it was significantly reduced after 3 days of exposure (Villafane *et al.* 2000). The differential sensitivity of pigments to UVR has also been studied for MPB organisms. Phycobilins of cyanobacterial mats appear to be more sensitive than chlorophylls and carotenoids, the latter often increasing at UV-B exposure (Garcia-Pichel & Castenholz 1994, Quesada *et al.* 1995, Quesada & Vincent 1997). However, in experiments conducted on intact sediment communities dominated by diatoms, no changes in pigment composition (expressed as ratios to chl a) were observed (Sundback *et al.* 1996, 1997, Odmark *et al.* 1998, Wulff *et al.* 1999, 2000), with one exception: higher carotenoid content was observed at enhanced UV-B levels in a Gyrosigma mat, probably reflecting a UV-B - protecting strategy (Underwood *et al.* 1999).

In marine macroalgae, various responses were also found when addressing the effects of UVR upon various pigments. Exposure of Porphyra umbilicalis to artificial UVR levels decreased chl a and phycocyanin concentrations by 65 and 67 per cent, respectively, whereas carotenoids and phycoerythrin decreased by as much as 75 and 82 per cent, respectively (Aguilera *et al.* 1999).

Furthermore, and under ambient levels, UVR not only decreased the concentration of chl a and biliproteins in the red alga Porphyra leucosticta, but the pattern of daily variation was also affected (Figueroa *et al.* 1997b). The damage of photosynthetic pigments by UVR in *P. leucosticta* was suggested to be the cause of a decrease in photosynthetic rates. However, in Macrocystis pyrifera, it was found that the main light-harvesting complex of this alga, the fucoxanthin - chlorophyll protein, was the specific site for UV damage (Wood 1989). Finally, in Ulva rigida (Altamirano *et al.* 1999) and Dasycladus vermicularis (Perez-Rodriguez *et al.* 1998), the content of chlorophyll and carotenoids was significantly higher in the presence of UV-B than that in the control (PAR - only), suggesting the presence of an efficient protective – pigment mechanism.

Cell Morphology and Size

When evaluating the photosynthetic responses to UVR of diverse organisms, some studies have revealed the importance of cell size (Karentz *et al.* 1991a, Helbling *et al.* 1992b, 2001a, b, Laurion & Vincent 1998). In phytoplanktonic organisms, it was found that, although there is certainly variability in responses, small cells – with a relatively high surface to volume ratio – are more resistant to photosynthesis inhibition but more vulnerable to DNA damage (Buma *et al.* 2001b, Helbling *et al.* 2001a, b). On the other hand, and provided that microplanktonic cells (20-200 µm) do not have high concentrations of UV - absorbing compounds, they are more vulnerable to UVR (in terms of photosynthesis). This has been demonstrated in a comparative study carried out in the Andean lakes (Helbling *et al.* 2001b), where it was found that larger phytoplanktonic cells had a higher kinetics of inhibition and hence were more affected by UVR than smaller cells.

For MPB organisms, on the other hand, there are

contradicting findings as to whether UV-B - related changes in species composition are related to cell size or are due to taxon—specific sensitivity (Garcia-Pichel 1994, Peletier *et al.* 1996, Helbling *et al.* 2001b). As seen for planktonic algae, increasing size may occur both on an individual species level, as cell division is hampered (Karentz *et al.* 1991a, Behrenfeld *et al.* 1992), and on community level, as species with larger cell-size could be favoured (Garcia-Pichel 1994, Wangberg *et al.* 1996). For MPB there is some indication for the latter, but not for the former. Bothwell *et al.* (1993) found that large, stalked diatom species increased their dominance at UV-B exposure during periphyton succession, and Vinebrook & Leavitt (1996) found that the growth of the small - sized diatom Achanthes minutissima was suppressed under UV-B exposure.

Besides size, the morphology also seems to influence the response of algae to solar radiation. This has been shown particularly for macroalgae. A comparison between the red algae Porphyra leucosticta and Rissoella verruculosa, which have comparable zonation patterns at intertidal sites, shows the different photo - protective strategies of these algae (Figueroa *et al.* 1997b, Flores-Moya *et al.* 1998). This is probably related to different absorption properties because of the thallus thickness and pigment composition. *P. leucosticta* has a thin thallus consisting of one cell layer in which light transmits rapidly and homogenously towards the harvesting complexes, whereas in R. verruculosa, which has a more complex structure, some scattering of photons through the multilayered thallus (self-shading) may take place.

This was evident when the algae were exposed to full solar radiation, and in Porphyra UVR accounted for about 30 per cent of the total photoinhibition, whereas no effects were observed in Rissoella. In addition, some studies hint about the importance of different life stages, which are closely related with size. Although studies have focused on the macrothallus or adult stages, it is expected that UVR stress would be more evident in the microscopic life stages (single - and few - celled), mainly due to their structural simplicity. These studies, in addition, bring about important consequences for algal zonation.

For example, depth distribution patterns of large kelps have been frequently thought to reflect the light requirements of establishment stages (spores, embryos, etc.). Paradoxically, most studies performed to address the relationship between the physiological performance under different light environments have been done with the large sporophytes, whereas spore adaptation has been largely overlooked (Huovinen *et al.* 2000). The question of whether early developmental stages of macroalgae, particularly spores, are more susceptible to UVR than larger life history phases, has been less addressed (Dring *et al.* 1996a).

If this is so, it is reasonable to think that the physiological adaptation of spore stages (such as the ability to acclimate to different light climates) will have consequences for the whole population dynamics (Reed 1990). UV can reduce photosynthetic rates of both micro and macroalgae by direct effects on the photosynthetic apparatus through (1) pigment photobleaching in the photosynthetic antenna, (2) reduction of proteins in photosystem II, (3) decrease of enzyme activity in the Calvin cycle, and (4) inhibition of carbonic anhydrase activity, as well as via indirect effects, such as DNA damage. Few studies have analysed the effects of UVR on photosynthetic responses other than carbon assimilation, for example nitrogen or phosphorus assimilation. Thus, analysing the effects of UVR on such integrated metabolic processes should become an experimental effort of high priority. A fact that complicates the study of UVR effects on photosynthetic organisms is that ozone depletion is occurring parallel to other global environmental changes, such as the increase in CO_2 and temperature, as well as the increasing eutrophication and acidification of natural waters. In order to predict the effects of increased UV-B radiation on aquatic ecosystems it is necessary to take into account the changes in other environmental factors.

The increase in UV-B levels does affect algal physiology and ecology, including biogeochemical cycles in the coastal zone and enhanced radiation may have a significant global - scale climatic impact (Kelly 1986, Williamson & Zagarese 2003, Zepp 2003). Changes in productivity or diversity of aquatic primary producers due to elevated UV-B levels are likely to bring about alterations on several trophic levels of coastal marine food webs. Therefore, changes in community structure and ecosystem function can be expected. Even ambient UVR levels can have a significant effect on benthic and water-column algal communities.

For example, a general feature of MPB response is that ambient UV-B levels can exert a selective pressure on early successional stages on both hard and soft substrata. However, systems appear to vary greatly in susceptibility, depending on climate, the availability of refuges and nutrients, as well as the level of productivity and structure of the food webs. The local ecological implication of the initial selective pressure during early colonisation will thus depend on the general importance of colonisation events in relation to UV-B exposure. In the case of macroalgae, UV-B may also function as a selective pressure at the time of early colonisation and recruitment. Consequently, macroalgal zonation can be determined by the different resistance against UV-B radiation of spore germination and growth of young plants (Huovinen *et al.* 2000).

How Plants Cope with the Desert Climate ?

Desert plants tend to look very different from plants native to other regions. They are often swollen, spiny, and have tiny leaves that are rarely bright green. Their strange appearance is a result of their remarkable adaptations to the challenges of the desert climate. Aridity is the sole factor that defines a desert and is the primary limitation to which desert organisms must adapt.

Desert plants have developed three main adaptive strategies: succulence, drought tolerance and drought avoidance. Each of these is a different but effective suite of adaptations for prospering under conditions that would kill plants from other regions.

Succulence

Succulent plants store water in fleshy leaves, stems or roots. All cacti are succulents, as are such non-cactus desert dwellers as agave, aloe, elephant trees, and many euphorbias. Several other adaptations are essential for the water storing habit to be effective.

A succulent must be able to absorb large quantities of water in short periods. Desert rains are often light and brief, and the soil dries rapidly under an intense sun. To cope with these conditions, nearly all succulents have extensive, shallow root systems. The roots of a saguaro extend horizontally about as far as the plant is tall but are rarely more than four inches (10 cm) deep. The water-absorbing roots are mostly within the upper half inch (1.3 cm).

Succulents must be able to maintain their water hoards in a desiccating environment and use it as efficiently as possible. The stems and leaves of most species have waxy cuticles that render them nearly waterproof when the stomates are closed. Water is further conserved by reduced surface areas; most succulents have few leaves (agaves), no leaves (most cacti), or leaves that are deciduous in dry seasons (elephant trees, ocotillos, boojums).

Many succulents, as well as semi-succulents such as most yuccas, epiphytic orchids, and xerophytic bromeliads, possess a water-efficient variant of photosynthesis called CAM, an acronym for Crassulacean Acid Metabolism. CAM plants open their stomates for gas exchange at night and store carbon dioxide. By day, while the stomates are closed, photosynthesis is conducted using the stored carbon dioxide. Because of the lower temperatures and higher humidity at night, CAM plants lose one-tenth as much water per unit of carbohydrate synthesised as standard C_3 plants.

Another valuable attribute of CAM plants is their capability for idling metabolism during droughts. When CAM plants become water-stressed, the stomates remain closed both day and night; gas exchange and water loss nearly cease. The plant, however, maintains a low level of metabolism in the still-moist tissues. Just as an idling engine can rev up to full speed more quickly than a cold one, an idling CAM plant can resume full growth in 24 to 48 hours after a rain. Therefore, succulents can take rapid advantage of ephemeral surface moisture.

Stored water in an arid environment requires protection from thirsty animals. Most succulent plants are spiny or toxic, often both. Some protect themselves by growing only in inaccessible locations. Still others rely on camouflage. Arizona night blooming cereus, for example, closely resembles the dry stems of the shrubs in which it grows.

Drought Tolerance

Drought tolerance (or drought dormancy) refers to a plant's ability to withstand desiccation without dying. Plants in this category often shed leaves during dry periods and enter a deep dormancy. Most water loss is from transpiration through leaf surfaces, so dropping leaves conserves water in the stems. Some plants that do not normally shed their leaves have resinous coatings that retard water loss (e.g., creosote bush).

The roots of drought tolerant shrubs and trees are extensive compared to those of plants in wetter climates, covering an area up to twice the diameter of the canopy. They exploit the soil at greater depth than the roots of succulents; sometimes they extend to extreme depths (e.g., mesquite). Most of a mesquite's roots, however, are within three feet (0.9 m) of the surface.

Rooting depth controls opportunities for growth cycles. In contrast to the succulents' shallow-rooted strategy, a substantial rain is required to wet the deeper root zone of shrubs and trees. After a soaking rain has fallen, shrubs such as brittlebush and creosote take a few weeks to resume full growth from deep dormancy. The trade-off between this strategy and that of succulents is that once the deeper soil is wetted by several rains it stays moist much longer than the surface layer, supporting several weeks of growth.

Succulents can absorb water only when the soil is nearly saturated. In contrast drought tolerant plants can absorb water from soil that is much drier. Similarly these plants can photosynthesise with low leaf moisture contents that would prove fatal to most plants.

Drought Avoidance

Annual plants escape unfavourable conditions by not existing. They mature in a single season, then die after channelling all of their life energy into producing seeds instead of reserving some for continued survival.

Most Sonoran Desert annuals will germinate only during a narrow window in the fall, after summer heat has waned and before winter cold arrives. During this window of opportunity there must be a soaking rain of at least one inch for most species. This combination of requirements is survival insurance: an inch of rain in the mild weather of fall will provide enough soil moisture that the germinating seeds will probably mature and produce seeds even if almost no more rain falls in that season. There is still further insurance: even under the best conditions not all of the seeds will germinate; some remain dormant. Although the mechanisms are not known, a percentage of any year's crop of desert lupine seeds will not germinate until they are ten years old.

Seedlings rapidly produce rosettes of leaves during the mild fall weather, remain flat against the ground as they grow more slowly through the winter, and bolt into flower in the spring. Since the plants are inconspicuous until they begin the spring bolt, many people mistakenly think that spring rains produce our wildflower displays.

Annuals are common only in communities that have dry

seasons, where the spacing of perennial plants is determined by the rooting space required to obtain enough moisture to survive the driest years. In the occasional wetter years both open space and moisture are available to be exploited by a population of fast-growing annuals. The more arid the habitat, the greater the proportion of annual species. Half of the Sonoran Desert's flora is comprised of annual species. In the driest habitats up to 90 per cent of the plants are annuals.

The desert environment may seem hostile, but this is purely an outsider's viewpoint. Adaptations enable indigenous plants and animals not merely to survive here, but to thrive most of the time.

Bibliography

Arora, D.K.: *Handbook of Applied Mycology*, Dekker, London, 1992.

Aslam, M.S.: *Handbook of Biology*, Campus Books, New Delhi, 2000.

Aslam, M.S.: *Immunobiology*, Campus Books, New Delhi, 2000.

Bay, E.G.: *Science: An Introduction*, Westview Press, Westview Special Series on Africa, London, 1982.

Board, R.G.: *A Modern Introduction to Food Microbiology*, Blackwell Scientific Publications, London, 1983.

Cherry, J.P.: *Protein Functionality in Foods*, American Chemical Society, Washington D.C., 1981.

Dumazedier, J.: *Towards a Society of Leisure*, Free Press, New York, 1967.

Duncan, W.A.: *Engendering School Learning: Science, Attitudes and Achievement among Girls and Boys in Botswana*, University of Stockholm, Stockholm, 1989.

Dyke, S .F.: *The Chemistry of Vitamins*, Interscience Publishers, New York, 1965.

Eley, G.: *Wild Fruits and Nuts*, E.P. Publishing Ltd., Yorkshire, 1976.

Gill, L.S.: *Science, Technology and Gender*, Polity Press in Association with the Open University, Cambridge, London, 1992.

Govindjee, Šesták Z. and Peters W.R.: *The Early History of Photosynthetica*, Comenius University, Bratislava, 2002.

Haisel, D. and Pospíšilová, J.: *The Effect of Water Stress on Photosynthetic Pigments Contents and Gas Exchange Parameters*, Comenius University, Bratislava, 2004.

Hudson, F.: *Development in Food Proteins*, Applied Science Publisher, London, 1982.

Kak, Subhash C.: *Indian Physics: Outline of Early History*, ArXiv, London, 2003.

Krause, M.V. and L.K. Mahann: *Food Nutrition and Diet Therapy*, W.B. Saunders Co., Philadelphia, 1979.

Malcolm, S.: *Science and Technology for People*, Tamil Nader State Council for Science and Technology, Madras, 1985.

Samba, Murthy K.V.: *Science and Technology in Medieval India*, Pratap, Agra, 1982.

Šesták, Z.: *Bibliography of Reviews and Methods of Photosynthesis*, Kluwer Academic Publishers, London, 1998.

Šesták, Z.: *Chlorophyll Fluorescence Kinetics Depends on Age of Leaves and Plants*, Kluwer Academic Publishers, London, 1999.

Sharma, P.V.: *History of Medicine in India*, Academic, New Delhi, 1992.

Singh, Harkishan: *Pharmaceutical Education*, Vallabh Prakashan, New Delhi, 1998.

Sponsors: *Proceedings of the Southeast Asian Seminar on Women and Science in Developing Countries*, Kovalevskaia Fund, New York, 1987.

Stolte-Heiskanen, V.: *Science: Role of Changing India*, ISSC, New Delhi, 1991.

Subbarayappa, B.V.: *Science and Technological Exchanges between India and Soviet Central Asia*, Educational, New Delhi, 1985.

Theresa, Montini: *Science: Technology and Human Values*, Oxford, New York, 1993.

Turnbull, David, and Helen Watson-Verran: *Handbook on Science, Technology and Society*, Sage, Beverly Hills, London, 1994.

Ulman, P., Èatský J. and Pospíšilová, J.: *Photosynthetic Traits in Wheat Grown under Decreased and Increased CO_2 Concentration after Transfer to Natural CO_2 Concentration*, Institute of Plant Molecular Biology, Èeské Budìjovice, 2000.

Vágner, M., Lipavská H. and Tichá I.: *Photosynthetic Pigments and Gas Exchange of in Vitro Grown Tobacco Plants as Affected by CO_2 Supply*, International Society of Photosynthesis, Montreal, 1999.

Valcke, R.: *Transgenic Pssu-ipt Tobacco under Biotic Stress*, CSIRO Publishing, Melbourne, 2001.

Van der Est, A. and Bruce, D.: *Photosynthesis: Fundamental Aspects to Global Perspectives*, International Society of Photosynthesis, Montreal, 2005.

Veldhnis, M.K.: *Food Science and Technology*, The AVI Publishing Co., Connecticut, 1977.

Vièánková, A., Kutík J., Holá D., Koèová M. and Wilhelmová N.: *The Chloroplast Ultrastructure and Photosynthetic Characteristics of Maize Leaves in Various Parts of the Leaf Blade and during the Diurnal Cycle*, Institute of Plant Molecular Biology, Èeské Budìjovice, 2001.

Vijaya Lakshmi, Sharma: *Science in Pre-Modern India*, D.K. Print World, New Delhi, 2001.

Vomáèka, L. and Pospíšilová J.: *Rehydration of Sugar Beet Plants after Water Stress: Effects of Cytokinins*, Institute of Plant Molecular Biology, Èeské Budìjovice, 2003.

Vomáèka, L.: *Gas Exchange and Cytokinins*, CSIRO Publishing, Melbourne 2001.

Wardha, B.: *Science and Technology in India*, Centre of Science for Villages, Wardha, 1983.

Whitaker, J.R.: *Principles of Enzymology for Food Science*, Marcel Dekker, New York, 1977.

Winslow, A.: *Science: History*, Duke University Press, London, 1990.

Zima, M.: *Ecophysiology of Plant Production Processes in Stress Conditions*, Slovak Agr. Univ., Nitra, 2000.

Zimmerman, J.: *Science in India: Today and Tomorrow*, Praeger Publishers, London, 1983.

Index

❑❑❑